# STEM Education in the Post-Pandemic Learning Space

# Africa in the Global Space

Edited by Edward Shizha

Vol. 11

Brantina Chirinda and Jayaluxmi Naidoo (eds.)

# STEM Education in the Post-Pandemic Learning Space

Digitilization in Africa

**PETER LANG**

New York · Berlin · Bruxelles · Chennai · Lausanne · Oxford

**Bibliographic Information published by the Deutsche Nationalbibliothek**
The Deutsche Nationalbibliothek lists this publication in the Deutsche Nationalbibliografie; detailed bibliographic data is available online at http://dnb.d-nb.de.

Library of Congress Control Number: 2024058970

Cover image: © Waeruslan Waedaraseh. (2024). Vector illustration of Africa continent map. iStock. https://www.istockphoto.com

ISSN 2576-358X
ISBN 978-3-0343-5239-0 (Print)
ISBN 978-3-0343-5240-6 (E-PDF)
ISBN 978-3-0343-5241-3 (E-PUB)
DOI 10.3726/b22671

© 2026 Peter Lang Group AG, Lausanne (Switzerland)
Published by Peter Lang Publishing Inc., New York (USA)

info@peterlang.com

All rights reserved.
All parts of this publication are protected by copyright.
Any utilization outside the strict limits of the copyright law, without the permission of the publisher, is forbidden and liable to prosecution. This applies in particular to reproductions, translations, microfilming, and storage and processing in electronic retrieval systems.

This publication has been peer reviewed.

www.peterlang.com

# Table of Contents

List of Figures ... vii
List of Tables ... ix
Foreword ... xi
Preface ... xv

CHAPTER 1
STEM Education in the Post-Pandemic Learning Space ... 1
*Jayaluxmi Naidoo and Brantina Chirinda*

CHAPTER 2
Curriculum (Re)Configuration and Transformative
Pedagogies for Effective Science, Technology, Engineering,
and Mathematics Education in Africa ... 7
*Crispen Bhukuvhani, Alois Solomon Chiromo, Charles Chikunda,
and Locadia Bhukuvhani*

CHAPTER 3
Digital Transformation in Science, Technology,
Engineering, and Mathematics Education in Sub-Saharan Africa ... 29
*Brantina Chirinda, Eddie M. Mulenga, and Gladys Sunzuma*

CHAPTER 4
Creating Digitalized and Virtual Spaces in Real-Time
Egyptian Post-COVID Mathematics Education Classrooms:
Insight from the Higher Education Sector in STEM ... 51
*Mariam Makramalla*

TABLE OF CONTENTS

**CHAPTER 5**
Strengths, Challenges, and Implications of Digital Pedagogy for Mathematics Education: Exploring South African Postgraduate Students' Experiences .......... 71
*Jayaluxmi Naidoo and Rajendran Govender*

**CHAPTER 6**
Teachers' Experiences of Blended Classrooms in Mathematics, Science, and Technology Academic Hubs .......... 87
*Zingiswa Jojo and Puleng Motseki*

**CHAPTER 7**
Postgraduate Students' Experiences of Online Teaching of Science, Technology, Engineering, and Mathematics Subjects .......... 107
*Asheena Singh-Pillay*

**CHAPTER 8**
Infusion of Technology in the Teaching and Learning of Mathematics in a South African University .......... 125
*Neliswa Gqoli*

**CHAPTER 9**
STEM Teaching and Learning in Early Childhood Classrooms During and Post the COVID-19 Pandemic in Zimbabwe .......... 139
*Agnes Pakombwele*

**CHAPTER 10**
STEM Education in the New Normal: Teacher Educators' Experience on the Use of Digital and Face-to-Face Pedagogy .......... 159
*Esther Kibga and Fredrick Mtenzi*

**CHAPTER 11**
Mathematics Teaching During and After Times of Crisis in the African Higher Education Context .......... 181
*Shonisani Agnes Mulovhedzi, Nosisi Nellie Feza, and Tawanda Runhare*

**CHAPTER 12**
Summing Up: Digitalization in STEM Education in Post-Pandemic Africa .......... 201
*Brantina Chirinda and Jayaluxmi Naidoo*

About the Editors .......... 211
Notes on Contributors .......... 213

# List of Figures

| | | |
|---|---|---|
| Figure 2.1: | Five-stage blended teaching and learning strategy for STEM | 14 |
| Figure 2.2: | The VALUES model | 19 |
| Figure 3.1: | Unified theory of acceptance and use of technology (UTAUT2) framework | 33 |
| Figure 4.1: | Egyptian K-12 sector distribution leading to higher education | 53 |
| Figure 4.2: | Sample for the study | 55 |
| Figure 4.3: | Inclusive pedagogy framework: conceptualization for this study | 57 |
| Figure 4.4: | Single-embedded case study design | 58 |
| Figure 4.5: | Sampling layout | 60 |
| Figure 4.6: | Analytical approach: inductive data coding | 62 |
| Figure 6.1: | The technological pedagogical content knowledge (TPACK) framework | 92 |
| Figure 7.1: | The components of the TPACK framework | 111 |
| Figure 8.1: | The technological pedagogical content knowledge model | 129 |
| Figure 10.1: | STEM education and its change | 163 |
| Figure 10.2: | Unified theory of acceptance and use of technology (UTAUT) model | 166 |
| Figure 11.1: | The improvement of teaching mathematics after a pandemic crisis in higher education | 189 |

# List of Tables

| | | |
|---|---|---|
| Table 2.1: | African Union, Agenda 2063—aspirations, goals, and priority areas | 9 |
| Table 3.1: | Zambian teachers' responses to mathematics teaching during the pandemic era | 38 |
| Table 4.1: | Abbreviations of the members of the sample | 60 |
| Table 4.2: | Open-ended interview structure | 61 |
| Table 4.3: | Example of inductive data coding | 63 |
| Table 5.1: | Pseudonyms used for the 22 participants who actively participated in the two discussion forums | 76 |
| Table 6.1: | Summary of Grade 9 mathematics performance in sub-Hub A | 89 |
| Table 7.1: | Discussion forum questions | 112 |
| Table 7.2: | Codes and themes | 113 |
| Table 9.1: | Demographic data of participants | 145 |
| Table 10.1: | Number of estimated Internet users and percentage | 160 |
| Table 10.2: | Number of online radios and TVs | 161 |

# Foreword

The COVID-19 pandemic has catalyzed a profound shift in the landscape of education globally, and nowhere is this more evident than in the fields of Science, Technology, Engineering, and Mathematics (STEM) education across Africa. This transformative period has exposed the vulnerabilities and opportunities inherent in our educational systems, particularly in integrating digital technologies into teaching and learning. The book, "STEM Education in the Post-Pandemic Learning Space: Digitalization in Africa," comprehensively explores these shifts, presenting cutting-edge research and insightful analyses from various African contexts. Chapter 2, "Curriculum (re)Configuration and Transformative Pedagogies for Effective Science, Technology, Engineering, and Mathematics Education in Africa," addresses the crucial need for a responsive and dynamic STEM curriculum. The African Union's Agenda 2063 emphasizes the importance of Science, Technology, and Innovation (STI) in sustainable development. This chapter underscores the necessity of an interdisciplinary, collaborative, and technology-enhanced curriculum that aligns with real-life challenges to advance sustainable development. It also proposes a blended learning model and the integration of Artificial Intelligence as innovative pathways to enhance STEM education in Africa.

In Chapter 3, titled "Digital Transformation in Science, Technology, Engineering, and Mathematics Education in Sub-Saharan Africa," the authors examine the immediate impact of the COVID-19 pandemic on STEM education in Southern Africa. The sudden shift to remote teaching revealed significant gaps in digital literacy among teachers and the digital divide affecting access to technology. Through qualitative multi-case studies from Zambia, South Africa, Namibia, and Zimbabwe, this chapter explores how these challenges were addressed and the role of digital tools in continuing STEM education during the crisis. The findings highlight the pandemic as a pivotal moment for reimagining STEM education with digital integration at its core. Also, the author of Chapter 4 researches "Creating Digitized and Virtual Spaces in Real-time Egyptian post-COVID Mathematics Education Classrooms: Insight from the Higher Education Sector in STEM," which offers a nuanced perspective on the inclusivity of online instruction. Explicitly focusing on female postgraduate students with mobility issues, this chapter demonstrates how online education can address inequities and support marginalized groups. The inclusive pedagogy framework used in this study provides valuable lessons for making higher education more accessible and equitable in the post-pandemic era. Chapter 5, "Strengths, Challenges, and Implications of Digital Pedagogy for Mathematics Education: Exploring South African Postgraduate Students' Experiences," continues the exploration of digital pedagogy. This chapter utilizes the SAMR model to evaluate the integration of digital tools in mathematics education. The experiences of postgraduate students shed light on the potential and obstacles of digital pedagogy. Key insights include the necessity of preparation workshops and the significance of digital resources and collaborative engagement for effective learning.

Chapter 6, "Teachers' Experiences of Blended Classrooms in Mathematics, Science, and Technology Academic Hubs," centers on the Mpumalanga Province in South Africa. The chapter investigates the implementation of blended learning and its impact on teaching practices during and after the pandemic. The findings underscore the significance of collaboration, technical support, and the availability of e-learning materials for successful blended learning environments. Meanwhile, Chapter 7 delved into the "Postgraduate Students' Experiences of Online Teaching of Science, Technology, Engineering, and Mathematics Subjects" addresses the complexities of teaching STEM subjects online in a linguistically diverse environment. The authors explore the use of ICT for translanguaging and linking crosscutting concepts, which reveals the opportunities and challenges teachers face. This chapter provides practical insights into leveraging ICT to enhance STEM education in developing countries.

Chapter 8, "Infusion of Technology in the Teaching and Learning of Mathematics in a South African University," focuses on integrating technology in a rural university in the Eastern Cape Province, South Africa. Using the TPACK framework, the study examines lecturers' challenges and positive attitudes towards technology use. The findings highlight the importance of professional development and infrastructure enhancements to support effective digital learning in rural contexts.

Chapter 9, "STEM Teaching and Learning in Early Childhood Classrooms During and Post the COVID-19 Pandemic in Zimbabwe," shifts the focus to early childhood education. The study reveals the significant disruptions caused by the pandemic and the subsequent challenges in adapting for teachers and learners. Recommendations include implementing long-term policies and training to better equip educators for future disruptions. Chapter 10 explores the lived experiences of educators in Tanzania in "STEM Education in the New Normal: Teacher Educators' Experience on the Use of Digital and Face-To-Face Pedagogy." This chapter emphasizes educators' readiness to adapt to different learning environments and highlights the importance of regular professional development to enhance digital and face-to-face pedagogy.

Chapter 11, "Mathematics Teaching During and After Times of Crisis in the African Higher Education Context," investigates how lecturers and student teachers coped with the pandemic. This chapter highlights the importance of blended pedagogies and provides recommendations on structuring mathematics activities in the Foundation Phase to meet the demands of the fourth industrial revolution (4IR). Finally, in Chapter 12, "Summing Up: Digitalization in STEM Education in Post-Pandemic Africa," the edited book concludes by examining the primary themes presented, summarizing the critical points of each chapter, providing final reflections and thoughts, sharing personal insights and lessons learned, and discuss broader implications of the book.

This book demonstrates the resilience and innovation in African STEM education during unprecedented change. Moving forward, these chapters provide valuable insights and practical recommendations for teachers, policymakers, and researchers committed to advancing STEM education in a post-pandemic world.

Musa Adekunle Ayanwale
Department of Science and Technology Education,
University of Johannesburg,
South Africa

# Preface

Digitalization in Science, Technology, Engineering, and Mathematics (STEM) education in Africa post-pandemic holds great potential for improving the quality and accessibility of STEM learning. The COVID-19 pandemic underscored the importance of digital tools and online educational resources. Digitalization in STEM education in Africa post-pandemic has gained momentum as countries and institutions seek to leverage technology to overcome educational disruptions, improve access to quality STEM education, and prepare students for careers in the digital age.

During the COVID-19 pandemic, STEM teachers and teacher educators at universities across Africa abruptly found their classrooms and courses transitioned to completely online contexts. Most STEM teachers and teacher educators in Africa had to adapt to online teaching for the first time without resources, knowledge, or support. The COVID-19 pandemic drove African countries to transition from face-to-face STEM classrooms to virtual classrooms. Now, close to four years later, the time for reflection is here. What have STEM teachers learned from this extraordinary era regarding the teaching and learning of STEM? What have STEM teacher educators learnt from this extraordinary era regarding higher education pedagogy? How can STEM teachers and teacher educators integrate the rewarding features of online teaching and learning experienced during the COVID-19 pandemic into physical classrooms? What does the new normal look like? How can STEM teachers and teacher educators incorporate the practices of both face-to-face and online teaching?

This volume brings together STEM education researchers from different parts of Africa. It seeks to explore and answer these questions with reference to STEM teaching and learning in classrooms and universities across Africa. From the work presented in this volume, we hope to provide conclusive solutions on how STEM teachers and teacher educators around the globe can design their classrooms around blended, hybrid, or hybrid flexible (hyflex) pedagogies that combine face-to-face teaching and online instruction.

In Africa, the pandemic exacerbated pre-existing inequalities and the digital divide in education. Some chapters in the book address these issues and explain how the innovations explored during the COVID-19 pandemic promoted digital transformation in African STEM education.

This book is timely and of interest to STEM teacher educators, teachers, researchers, professional development providers, and curriculum designers.

Brantina Chirinda<br>
Johannesburg, South Africa<br>
<br>
Jayaluxmi Naidoo<br>
Durban, South Africa

CHAPTER 1

# STEM Education in the Post-Pandemic Learning Space

Jayaluxmi Naidoo[1] and Brantina Chirinda[2]
[1]University of KwaZulu-Natal
[2]University of Johannesburg

## Introduction

The post-pandemic era has substantially transformed Science, Technology, Engineering, and Mathematics (STEM) education globally. In Africa, the promotion of digitalization has resulted in a distinctive landscape for advancing STEM education (Maarman, 2023). These advancements include adopting online teaching and learning platforms, encouraging blended teaching and learning models, increasing accessibility, promoting curriculum adaptations, and increasing teacher professional development, training, and support.

With these advancements, it is important to consider both the challenges and opportunities. Some challenges include the inequality in access to tools, resources, and infrastructure that promote the digital divide (Kuteesa et al., 2024; Martens et al., 2020; Sithomola, 2021). In addition, teacher preparedness to adapt to digital-based pedagogy is also of concern (Naidoo and Singh-Pillay, 2022; Rwodzi and De Jager, 2021). The opportunities created by the advancement include economic growth, enhanced and interactive teaching and learning experiences, and global collaboration. With digitalization, we can potentially reach a larger and more diverse student population, including those in remote or underserved areas where physical and educational resources are scarce. This is transformative and makes STEM subjects more accessible and appealing to students.

Thus, while the post-pandemic era has advanced digitalization in STEM education in Africa, by addressing the challenges and embracing the opportunities, African countries can utilize the full potential of digitalization to enhance STEM education to improve educational outcomes, promote innovation, and increase economic growth. This volume explores familiar problems in a new way and delves into previously understudied issues concerning STEM education in Africa within the post-pandemic learning environment. An example is the VALUES model (refer to Figure 2.2) presented in Chapter 2 by Bhukuvhani et al. This model provides ways to incorporate African cultural, intellectual capital, and practices into STEM education. In Chapter 4, Mariam Makramalla challenges the mainstream opinion favoring in-person instruction for student learning of mathematical pedagogical practice tools. She does this by using an inclusive pedagogy framework and shedding light on the voices of female postgraduate students who would be marginalized in the traditional setup. The study undertaken by Asheena Singh-Pillay on the language of learning and instruction, digitalization, and STEM education in Chapter 7 is another understudied issue. Singh-Pillay argues that digitalization can address language challenges learners and teachers encounter with STEM subjects and promote disciplinary science understanding by linking crosscutting concepts.

## STEM Education in Africa

The post-pandemic era has advanced significant changes, motivated principally by the need for digital transformation. This has led to debates about the rural versus urban divide since there is a substantial gap in STEM education between rural and urban areas (Barakabitze et al., 2019; Mutambara and Bayaga, 2021). For example, schools in urban areas generally have better resources and access to laboratories than rural areas, which often lack basic resources and infrastructure. In Chapter 3 of this volume, Chirinda et al. note that it is crucial for policymakers and all stakeholders to address this gap and work towards providing adequate resources and opportunities for students in rural areas to ensure they have equal access to quality STEM education. Chirinda et al. stress that by acknowledging this issue, we can begin to implement strategies that will help to bridge the gap and create a more equitable educational system for all students, regardless of where they live.

Moreover, there exist gender disparities in that girls are under-represented in STEM domains as a result of social, cultural, and economic barriers (Founou et al., 2023; Mare, 2021). Additionally, many teachers in rural areas lack the necessary professional development, training, and support in STEM subjects compared to their urban counterparts (Maina et al., 2021; Wright, 2019). Professional development, training, and support in STEM subjects are important to improving teaching quality. In Chapter 8 of this volume, Nelisa Gqoli recommends that technology professional development in mathematics should be offered to teachers at rural universities. Online training modules and continuous professional development platforms for teachers can improve the quality of STEM education. Digital platforms can offer teachers up-to-date content, teaching resources, and peer support networks.

Furthermore, STEM curricula must be updated regularly to include new advancements and align with relevant local contexts and industries. As is evident, STEM teaching and learning in Africa is critical, with digitalization offering challenges and opportunities (Oladele et al., 2022). African nations can develop dynamic STEM education structures by enhancing teacher professional development, training, and support across all contexts, addressing infrastructural deficiencies, and promoting inclusive practices. These improvements are important for individual empowerment, encouraging sustainable development, and promoting African economic growth.

## Indigenous Knowledge Systems, Digitalization and Other Approaches for STEM Education in Africa

In the context of STEM education in Africa, Indigenous Knowledge Systems (IKS), digitalization, and other innovative approaches are important in establishing a more effective, relevant, holistic, and inclusive teaching and learning experience. Integrating IKS, digital tools, and other teaching methodologies can bridge cultural gaps, promote relevant contexts, encourage engaging learning, and enhance the practical application of scientific concepts (McKnight, 2024; Sitsha, 2023). Embracing IKS in STEM education enriches the learning experience by incorporating traditional wisdom and practices, which helps students appreciate diverse perspectives, develop critical thinking skills, and understand the relevance of STEM subjects in real-world contexts.

When considering integrating IKS in STEM education, traditional knowledge and practices of learners and their communities are highlighted. This ensures that learning is culturally relevant while considering rich local knowledge and traditions (Chahine, 2022; Hlalele, 2019). In addition, many IKS practices are fundamentally practical and sustainable. Thus, using IKS in STEM education encourages the use of sustainable practices and local problem-solving techniques in the classroom. Thereby creating a platform for contextual teaching and learning by using local examples. In this way, learners can link scientific concepts with their daily lives (Sitsha, 2023). Consequently, culturally relevant and practical examples can make STEM subjects more engaging and relatable for students.

Using technology, digital tools, virtual laboratories, and online platforms provides learners and teachers with access to various resources, simulations, and interactive content that support and enhances STEM teaching and learning. Virtual laboratories allow teachers and learners to practice scientific techniques and experiments in a virtual environment, which is useful for schools with limited resources and laboratory equipment (Byukusenge et al., 2022). In this way, learners are exposed to practical demonstrations and simulations regardless of their limited laboratory resources.

Incorporating other teaching methodologies, such as problem-based learning (PBL), can encourage students to consider projects exploring local challenges using scientific methods (Hernández-Ramos et al., 2021; Nagarajan and Overton, 2019). This approach enhances teaching and learning and encourages critical thinking. PBL makes learning relevant and engaging and helps develop important scientific skills, such as critical thinking and problem-solving. Using interdisciplinary projects also allows learners to develop a holistic understanding and collaborative skills. Moreover, encouraging group work and collaborative projects support learners to develop teamwork skills as they learn from their peers (Fang et al., 2021; Shofiyah et al., 2022).

Integrating IKS, digitalization, and other approaches into STEM education in Africa offers a favorable path towards a more inclusive, relevant, and effective educational experience (Ferreira-Meyers and Dhakulkar, 2021; Sitsha, 2023). These approaches can lead to developing sustainable solutions to meet local needs and conditions. By respecting and incorporating traditional knowledge, promoting the use of technology, and encouraging collaborative and PBL, teachers can promote learner engagement and learning outcomes (Gumartifa et al., 2023; Kassymova, 2020). These approaches can contribute to developing a skilled and advanced workforce to advance sustainable growth in Africa.

## Bibliography

Barakabitze, A. A., William-Andey Lazaro, A., Ainea, N., Mkwizu, M. H., Maziku, H., Matofali, A. X., …, & Sanga, C. (2019). Transforming African education systems in science, technology, engineering, and mathematics (STEM) using ICTs: Challenges and opportunities. *Education Research International*, 1–29.

Byukusenge, C., Nsanganwimana, F., & Tarmo, A. P. (2022). Effectiveness of virtual laboratories in teaching and learning biology: A review of literature. *International Journal of Learning, Teaching and Educational Research 21*(6), 1–17.

Chahine, I. C. (2022). Infusing Indigenous Knowledge Systems (IKS) into teaching integrated STEM disciplines: An empirical project. *Journal of Indigenous Research 10*(2022), 1–11.

Fang, M., Jandigulov, A., Snezhko, Z., Volkov, L., & Dudnik, O. (2021). New technologies in educational solutions in the field of STEM: The use of online communication services to manage teamwork in project-based learning activities. *International Journal of Emerging Technologies in Learning (iJET) 16*(24), 4–22.

Ferreira-Meyers, K., & Dhakulkar, A. (2021). Can open science offer solutions to science education in Africa? In D. Burgos & J. Olivier (Eds.), *Radical*

*Solutions for Education in Africa. Lecture Notes in Educational Technology* (pp. 149–174). Springer.

Founou, L. L., Yamba, K., Kouamou, V., Yeboah, E. E. A., Saidy, B., Jawara, L. A., ..., & Darboe, S. (2023). African women in science and development, bridging the gender gap. *World Development Perspectives* 31(2023), 1–4.

Gumartifa, A., Syahri, I., Siroj, R. A., Nurrahmi, M., & Yusof, N. (2023). Perception of teachers regarding problem-based learning and traditional method in the classroom learning innovation process. *Indonesian Journal on Learning and Advanced Education (IJOLAE)* 5(2), 151–166.

Hernández-Ramos, J., Pernaa, J., Cáceres-Jensen, L., & Rodríguez-Becerra, J. (2021). The effects of using socio-scientific issues and technology in problem-based learning: A systematic review. *Education Sciences* 11(10), 1–16.

Hlalele, D. J. (2019). Indigenous knowledge systems and sustainable learning in rural South Africa. *Australian and International Journal of Rural Education* 29(1), 88–100.

Kassymova, G., Akhmetova, A., Baibekova, M., Kalniyazova, A., Mazhinov, B., & Mussina, S. (2020). E-Learning environments and problem-based learning. *International Journal of Advanced Science and Technology* 29(7), 346–356.

Kuteesa, K. N., Akpuokwe, C. U., & Udeh, C. A. (2024). Theoretical perspectives on digital divide and ICT access: Comparative study of rural communities in Africa and the United States. *Computer Science & IT Research Journal* 5(4), 839–849.

Maarman, G. J. (2023). Basic sciences in higher education, and teaching approaches in the context of 21st-century advances: Time for a change? *South African Journal of Higher Education* 37(2), 132–150.

Maina, F., Smit, J., & Serwadda, A. (2021). Professional development for rural STEM teachers on data science and cybersecurity: A university and school districts' partnership. *Australian and International Journal of Rural Education*, 31(1), 30–41.

Martens, M., Hajibayova, L., Campana, K., Rinnert, G. C., Caniglia, J., Bakori, I. G., ..., & Oh, O. J. (2020). "Being on the wrong side of the digital divide": Seeking technological interventions for education in Northeast Nigeria. *Aslib Journal of Information Management* 72(6), 963–978.

Mare, A. (2021). Addressing digital and innovation gender divide: Perspectives from Zimbabwe. In C. Daniels, M. Dosso & J. Amadi-Echendu (Eds.), *Entrepreneurship, Technology Commercialisation, and Innovation Policy in Africa* (pp. 33–54). Springer.

McKnight, M. (2024). Investigating indigenous knowledge awareness among South African science teachers for developing a guideline. *Curriculum Perspectives 44*(1), 1–11.

Mutambara, D., & Bayaga, A. (2021). Determinants of mobile learning acceptance for STEM education in rural areas. *Computers & Education 160*, 1–16.

Nagarajan, S., & Overton, T. (2019). Promoting systems thinking using project- and problem-based learning. *Journal of Chemical Education 96*(12), 2901–2909.

Naidoo, J., & Singh-Pillay, A. (2022). Digital pedagogy for mathematics and technology education: Exploring the initiatives at one South African teacher education institution. In J. Olivier, A. Oojorah & W. Udhin (Eds.), Perspectives on teacher education in the digital age (pp. 223-241). Springer Nature.

Oladele, J. I., Ndlovu, M., & Ayanwale, M. A. (2022). Computer adaptive-based learning and assessment for enhancing STEM education in Africa: A fourth industrial revolution possibility. In B. Chirinda, K. Luneta, & A. Uworwabayeho (Eds.), *Mathematics Education in Africa: The Fourth Industrial Revolution* (pp. 131–144). Springer.

Olivier, A. Oojorah, & W. Udhin (Eds.), *Perspectives on Teacher Education in the Digital Age* (pp. 223–241). Springer Nature.

Rwodzi, C., & De Jager, L. (2021). Resilient English teachers' use of remote teaching and learning strategies in Gauteng resource-constrained township secondary Schools. *Perspectives in Education 39*(3), 62–78.

Shofiyah, N., Wulandari, F. E., Mauliana, M. I., & Pambayun, P. P. (2022). Teamwork skills assessment for STEM Project-Based Learning. *Journal Penelitian Pendidikan IPA 8*(3), 1425–1432.

Sithomola, T. (2021). The manifestation of dual socio-economic strata within the South African schooling system: A setback for congruous prospects of 4IR. *African Journal of Public Affairs 12*(3), 104–126.

Sitsha, M. (2023). *Exploring the integration of Indigenous Knowledge Systems (IKS) into the teaching of Life Sciences through Information and Communication Technologies (ICTs)* [Unpublished doctoral dissertation]. North-West University.

Wright, K. B. (2019). Improvement science as a promising alternative to barriers in improving STEM teacher quality through professional development. *The Clearing House: A Journal of Educational Strategies, Issues and Ideas 92*(1–2), 1–8.

CHAPTER 2

# Curriculum (Re)Configuration and Transformative Pedagogies for Effective Science, Technology, Engineering, and Mathematics Education in Africa

Crispen Bhukuvhani,[1] Alois Solomon Chiromo,[2]
Charles Chikunda,[3] and Locadia Bhukuvhani[4]
[1]Manicaland State University of Applied Sciences, Zimbabwe
[2]Midlands State University, Zimbabwe
[3]UNESCO Regional Office for Southern Africa
[4]Rubengera Teacher Training College, Rwanda

ABSTRACT
A relevant Science, Technology, Engineering, and Mathematics (STEM) curriculum responds adequately to global and societal challenges. African Union's Agenda 2063 outlines the continental commitment to Science, Technology, and Innovation (STI) to support sustainable development. STEM education is the anchor for industrialization and socio-economic sustainability for any country; hence, this places STEM learning areas and disciplines central to any country's curriculum. Therefore, for effective STEM teaching and learning, reorientation is required. The key characteristics of an effective STEM curriculum include interdisciplinarity, collaboration, authenticity, and technology-enhancement. These characteristics boost the construction of knowledge and skills in tackling real-life community issues and advance sustainable development. In line with the inevitable and ever-changing times of technology-supported learning, the proposed five-stage blended learning model in this chapter and the use of Artificial Intelligence may enhance learning for understanding and skills development in STEM disciplines. Taking cognizance of cultural ways of knowing amplifies the science and technology epistemological horizons. The VALUES model has been proposed to offer modalities of infusing African cultural intellectual capital and practices in STEM teaching and learning.

*Keywords:* STEM curriculum, sustainable development, transformative pedagogies, VALUES model

## Africa's Continental Commitment to STEM Education

The advancement and development of Africanized contextualized transformative curricula that speak to and address sustainable development in Africa has gained traction in the decolonization discourse. For example, to tackle and be prepared for perennial natural disasters (e.g., cyclones, droughts, and climate change) and epidemiological pandemics such as HIV/AIDS, Ebola, and COVID-19, among others that caught the continent unaware, African governments need to support curriculum development in education institutions in a concerted manner and create a common resource pool to support requisite human capital and requisite infrastructure development, taking advantage of Africa's rich and diverse intellectual capital and expertise.

Member countries from the various African sub-regional communities, such as the Economic Community of West African States (ECOWAS), the Southern

African Development Community (SADC) and the West African Community, advocate for the enactment of some commitments of special interest prioritized at their sub-regional levels. Typical examples include the Science, Technology, and Innovation Strategy for Africa 2024 and the SADC Protocol on Science, Technology, and Innovation, which are extrapolated from the African Union Agenda 2063. These work on operationalizing African Union Agenda 2063, Goal 2 on achieving a well-educated and skills-driven revolution that Science, Technology, and Innovation underpin at sub-regional levels. The sub-regional communities are responsible for engaging the governments of member states to develop or realign higher education curricula that seek to develop knowledge-driven and technology-anchored economies.

Responsible science can be a panacea for developing and promoting a peaceful Africa. The African Union acknowledges that the successful use of science, technology, and innovation is fundamental to sound policy-making, good governance, and industrial development (Royal Society, 2005). Science, technology, and innovation underpin attainment and sustainability regarding pertinent development issues in Africa, including poverty alleviation and economic growth. African Union, Agenda 2063—Aspirations, Goals and Priority Areas (Table 2.1) makes it abundantly evident that Africa prioritizes science, technology, and innovation as part of the solution to achieving sustainable development, with a particular focus on the goals and priority areas under Aspiration 1. This makes STEM education the cornerstone of Africa's development agenda.

On achieving or working towards achieving the African Union Agenda 2063, Aspiration 1—*Prosperous Africa, based on inclusive growth and Sustainable Development*, all the other aspirations will fall into place as they will be supported by a strong industrial and economic base anchored on sound scientific and technological footing.

Africa needs to have a harmonized STEM curriculum for human capital development for the skills, the continent requires to drive its Agenda 2063 to address the common developmental issues as a unified and diversified continent with talent that should be harvested for the common good and promotion of peace, cohesion, and sustainable development. For example, the Regional Universities Forum for Capacity Building in Agriculture (RUFORUM) is a consortium of universities in Africa. The coordinated approach to agriculture training, as a key STEM discipline, aims to improve the food and nutrition security situation on the continent towards zero hunger (SDG 1) and the eradication of poverty (SDG 2) in Africa. RUFORUM (2023) reports that a cooperation agreement with the AU supports the implementation of the Science, Technology, and Innovation Strategy (STISA-2024).

Table 2.1: African Union, Agenda 2063—aspirations, goals, and priority areas

| Aspirations | Goals | Priority Areas |
|---|---|---|
| 1) Prosperous Africa, based on Inclusive Growth and Sustainable Development. | (1) A high standard of living, quality of livelihood and well-being for all citizens. | · Incomes, jobs and decent work.<br>· Poverty, inequality and hunger.<br>· Social security and protection including persons with disabilities.<br>· Modern and liveable habitats. |
| | (2) Well-educated and skills revolution un derpinned by Science, Technology and Innovation. | · Education and STI skills driven revolution. |
| | (3) Healthy and well-nourished citizens. | · Health and nutrition. |
| | (4) Transformed economies. | · Sustainable and inclusive economic growth<br>· STI driven manufacturing/industrialisation and value addition.<br>· Economic diversification and resilience.<br>· Hospitality/Tourism |
| | (5) Modern Agriculture for increased productivity and production. | · Agricultural productivity and production. |
| | (6) Blue/green economy for accelerated economic growth. | · Marine resources and energy.<br>· Ports operations and marine transport. |
| | (7) Environmentally sustainable and climate resilient economies and communities. | · Sustainable natural resources and management and biodiversity conservation.<br>· Sustainable consumption and production patterns.<br>· Water security.<br>· Climate resilience and natural disasters preparedness and prevention.<br>· Renewable energy. |
| 2) Integrated Continent politically united and based on the ideals of Pan Africanism and the vision of African Renaissance. | (8) United Africa (Federal or Confederate) | · Framework and institutions for a united Africa. |
| | (9) Continental financial and monetary institutions are established and functional. | · Financial and monetary institutions |
| | (10) World class infrastructure across Africa. | · Communications and infrastructure connectivity. |

(Continued)

Table 2.1: (Continued)

| Aspirations | Goals | Priority Areas |
|---|---|---|
| 3) An Africa of Good Governance, Democracy, Respect for Human Rights, Justice and Rule of Law. | (11) Democratic values, practices, universal human rights, justice and the rule of law entrenched. | · Democracy and good governance.<br>· Human rights, justice and the rule of law. |
| | (12) Capable institutions and transformative leadership in place. | · Institutions and leadership.<br>· Participatory development and local governance. |
| | (13) Peace, security and stability is preserved. | · Maintenance and preservation of peace and security. |
| 4) A peaceful and Secure Africa. | (14) A stable and peaceful Africa. | · Institutional structure for AU instruments on peace and security. |
| | (15) A fully functional and operational APSA. | · Fully functional and operational APSA pillars. |
| 5) Africa with a strong cultural identity, common heritage, values and Ethics. | (16) African cultural Renaissance is pre-eminent. | · Values and ideals of Pan Africanism.<br>· Cultural values and African Renaissance.<br>· Cultural heritage, creative arts and business. |
| 6) Africa whose development is people driven, relying on the potential offered by African people, especially its women and youth, and caring for children. | (17) Full Gender Equality in all spheres of life. | · Women and girls empowerment<br>· Violence and discrimination against women and girls. |
| | (18) Engaged and empowered youth and children. | · Youth empowerment and children. |
| 7) Africa as a strong united, resilient and influential Global player and partner. | (19) Africa as a major partner in global affairs and peaceful co-existence. | · Africa's place in global affairs.<br>· Partnership. |
| | (20) Africa takes full responsibility for financing her development. | · African capital market.<br>· Fiscal system and public sector revenues.<br>· Development assistance. |

*Source:* African Union, 2023 https://au.int/en/agenda2063/goals

RUFORUM (2023) outlines that STISA-2024 seeks to accelerate Africa's transition to an innovation-led, knowledge-based economy, set to be achieved by:

1. improving STI readiness in Africa in terms of infrastructure, professional, and technical competence, and entrepreneurial capacity; and
2. implementing specific policies and programs in science, technology, and innovation that address societal needs holistically and sustainably (RUFORUM, 2023).

Another example of STEM education commitment is the setting up of Africa Centers of Excellence (ACEs). The World Bank (2023) reports that the ACEs project addresses higher-level skills development needs and innovative research requirements for the continent's priority development sectors in five main areas: science, technology, engineering, and mathematics (STEM); agriculture; health; environment; applied social science; and education. Since 2014, the program has supported over 80 centers in more than 50 universities throughout 20 countries. The program has thousands of students; more than one-third are females enrolled in postgraduate programs. The program is designed to meet worldwide standards for providing high-quality instruction and local specializations that support the continent's labor market demands. The ACEs are undertaking high-impact research addressing some of the region's most pressing development challenges, such as food insecurity, climate change, and infectious diseases (World Bank, 2023).

The African Union's Agenda 2063 aspirations and goals are related to the United Nations' Sustainable Development Goals (SDGs). STEM disciplines are crucial in addressing socio-economic ills in our communities, as pronounced in the AU's aspirations and the UN's SDGs. Therefore, a well-coordinated quality and inclusive STEM education (SDG 4) can positively influence the achievement of all the other SDGs. For instance, a strong STEM education footing yields agricultural and industrial productivity for enhancing food and nutrition security (SDG 1), eradication of poverty (SDG 2), promote good health and wellbeing (SDG 3), availability of clean water and sanitation facilities (SDG 5), use of clean energy (SDG 7), decent work (SDG 8), infrastructural development (SDG 9, 11, 12), responsible science promote peace (SDG 16), and international cooperation (SDG 17) for sustainable development.

Proper teaching of STEM disciplines that takes cognizance of their interdisciplinary and inter-related nature brings authentic learning linked to people's everyday experiential living, thereby making them relevant through the direct application of knowledge learnt in the classroom in real life or through seeing scenarios where science, technology, engineering, and mathematics concepts are applied in the real world.

## Nature and Characteristics of an Effective STEM Curriculum

According to Nite et al. (2017), an effective STEM curriculum principally has the following characteristics: interdisciplinary, hands-on, and problem-based. These key characteristics are explained below.

- *Interdisciplinary/Integrative:* Integrate the four STEM disciplines of science, technology, engineering, and mathematics. This helps students see how the disciplines are interconnected and how they can be used to solve real-world problems.
- *Hands-on Learning Experiences:* Provide students with opportunities for hands-on learning. This helps students develop a deeper understanding of STEM concepts and allows them to apply what they learn to real-world situations.
- *Problem-based:* Be problem-based. This involves giving students real-world problems to solve that require them to use STEM skills and knowledge. This helps students develop critical thinking, problem-solving, and collaboration skills.

In addition to these characteristics, an effective STEM curriculum should also be relevant to students' interests and everyday life experiences (Bhukuvhani, 2020), as these help students stay engaged in learning. A STEM curriculum should also be inquiry-based. In an inquiry-based curriculum, students actively participate in learning, asking questions, conducting experiments, and drawing conclusions. An inquiry-based STEM curriculum provides for the active construction of knowledge.

Linked to the hands-on characteristics covered above, STEM learning should be authentic. An authentic STEM curriculum means that students should work on problems that are relevant to the real world (Aikenhead, 1996, 2005; Aikenhead and Jegede, 1999). STEM teaching and learning provide for solving real-life socio-scientific issues through the application of scientific and technological knowledge.

An effective STEM education should be assessment-driven. The type of assessment should be in tandem with the desired and envisaged skills and competencies. The STEM curriculum should be designed to help students learn and grow by teaching them critical thinking, problem-solving, and collaboration skills essential for success in any field.

Here are some specific examples of how STEM education can be effective:

- Using the engineering design process to design and build a solar-powered car. This would allow students to apply their science, mathematics, and engineering knowledge to real-world problems.
- A mathematics class might use data analysis to investigate the impact of climate change on a local ecosystem. This would allow students to use their mathematics skills to solve real-world problems and make a difference in their community.
- An engineering class might use computer-aided design (CAD) software to create a new product prototype. This would allow students to use their engineering skills to solve a real-world problem and create something new.

STEM education develops skills and competencies such as problem-solving, critical thinking, and collaboration that prepare the populace for future jobs that may not currently exist. STEM jobs are in high demand and pay well, as they respond to the developmental needs of every society and are essential for success in any field.

## Adoption of Technology-Enhanced STEM Teaching and Learning

The adoption of technology-enhanced STEM teaching and learning has been growing in recent years. This is due to a number of factors, including the increasing availability of technology, the growing demand for STEM skills in the workforce, and the recognition that technology can be a powerful tool for engaging and motivating students. Lately, disasters such as COVID-19 have forced institutions of learning to ensure that teaching and learning cannot be interrupted. However, when technologies are deployed for teaching and learning, quality education should always be ensured.

There are a number of different ways that technology can be used to enhance STEM teaching and learning. Some of the most common methods include:

- ***Simulations:*** Simulations can be used to create immersive and realistic learning experiences for students. This can be especially helpful for STEM concepts that are difficult to visualize or experience in real life. For example, a science class may use a simulation to learn about the water cycle or an erupting volcano. The simulation allows students to see how the water cycle works or how a volcano erupts.

- **Games:** Games can be fun and engaging ways for students to learn STEM concepts. They can also help students develop critical thinking, problem-solving, and collaboration skills. For example, a mathematics class may use a dice game to learn about pobability. The game challenges students to solve probability problems to progress through the game.
- **Virtual Reality (VR):** According to Familoni and Onyebuchi (2024), VR creates immersive and interactive learning experiences. This can be especially helpful for STEM concepts that are difficult to see or experience in real life. For example, using VR in an engineering class to design a new product allows students to see their design in 3D and interact with it realistically.
- **3D Printing:** 3D printing can create physical models of STEM concepts, helping students visualize and understand them in a new way.
- **Data Analysis:** Data analysis tools can help students analyze and interpret data. This is an important skill for STEM students to develop, as many STEM careers involve working with data.

Al Hamad et al. (2024) note that adopting technology-enhanced STEM teaching and learning improves student learning outcomes. However, it is important to note that technology is not a silver bullet. Technology can be a powerful tool for learning, but it is not a replacement for good teaching (Bhukuvhani et al., 2012). Technology should be used in a way that is aligned with the learning goals and engaging and motivating for students.

From the discourse on providing an effective blend in the use of technology and physical laboratory sessions in teaching and learning STEM disciplines, a five-stage strategy proposition is presented in Figure 2.1. The five-stage strategy is outlined in detail regarding how technology-enhanced blended STEM teaching and learning could be implemented.

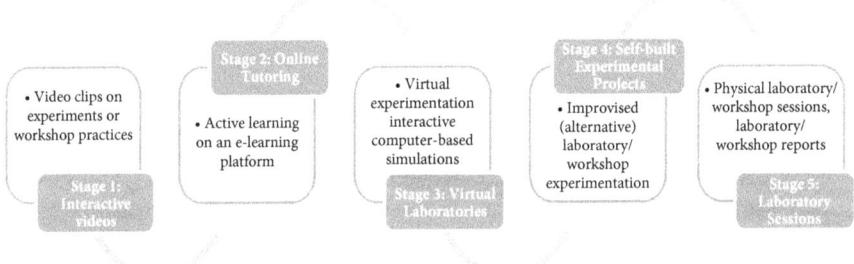

Figure 2.1: Five-stage blended teaching and learning strategy for STEM.

The stages of the five-stage blended teaching strategy for STEM disciplines are outlined as follows:

### Stage 1: Interactive videos
Students interface with interactive video clips of science experiments or engineering workshop practice sessions. According to Felder and Brent (2024), through interactive videos, students learn how to conduct science experimental activities, including procedures and safety precautions, and the scientific observations featured in the processes. These video clips will also motivate and increase the students' interest in learning science. The learning experience can be enhanced by introducing pop-up and multiple-choice questions that probe the students to think and test their understanding. In addition, multimedia has the power to animate and communicate dynamic information more accurately than a diagram. Visualizations help students mentally conceptualize phenomena that cannot be seen (Mekonnen et al., 2024).

### Stage 2: Online tutoring
Online tutoring using an e-learning platform can also help the learning process and hone students' critical thinking skills through active discussion. It enables tutors and learners to bring the face-to-face classroom into a virtual environment. Interactive, engaging online presence should be maintained. Such activities can generate new ideas and cultivate innovation (Lüy et al., 2024).

### Stage 3: Virtual laboratory experimentation/workshop practice
Virtual laboratories are multimedia applications that allow video and digital simulations of laboratory activities in a real manner. They do not have the risks and costs associated with laboratory experiments. Virtual laboratories contain simulated and remotely triggered laboratory experiments that allow students to learn the scientific concepts governing them.

It can be noted that stages 1–3 of this STEM teaching model may constitute the online learning phase of the blended learning mode. Blended learning combines face-to-face (stage 5) instruction and online learning. The blended learning mode of delivery offers several advantages for STEM education. Blended learning offers students to:

- Learn at their own pace. Students can watch online lectures and work on assignments at their convenience.
- Access resources from anywhere. Students can access online resources, such as simulations and games, from anywhere with an internet connection.
- Interact with other students from around the world. Students can participate in online discussions and collaborate on projects with students from other schools.

## Stage 4: Self-built experimental projects

According to Bhukuvhani et al. (2013), self-built experiments (SBE), which use common materials and equipment, have been suggested as an alternative to laboratory experiments. Kennepohl (2000) and MacQueen and Thomas (2009) suggest that these may be suitable for introductory science/technology courses. SBEs are forms of project-based learning approaches in STEM education. These use readily available materials, including the environment as living science and technology laboratories. For example, teaching the topic of ecosystems and biodiversity using the resources in a natural biosphere reserve.

Self-built experiments can be designed as follows:

*Step 1:* Formulate an experimental question/problem
*Step 2:* Set out the required materials, procedure, and safety precautions
*Step 3:* Select your experimental conditions (experimental conditions, variables manipulated and tested, etc.)
*Step 4:* Control for confounding factors (things that are NOT independent variables BUT can affect the outcome of the experiment)
*Step 5:* Define evaluation criteria

## Stage 5: Physical laboratory/workshop sessions

Students are required to perform experiments in the labs and are assessed on the quality of their lab reports. Under this model, students will only need to attend a single or a few practical session(s) (as may be required). The purpose of this or these session(s) will be to ascertain what the student has learned during the first four stages. A two-fold assessment system is proposed, e.g., a 3-h practical test where the student must submit a report at the end of the experiment(s), followed by a viva session.

This stage provides face-to-face (F2F) instruction to consolidate and reinforce already learned concepts through Stages 1–4 as part of the blended learning phase. Face-to-face instruction is the traditional mode of delivery for STEM teaching and learning, which involves students and teachers meeting in person in a classroom setting. The F2F delivery mode has many advantages for STEM education. These include the following:

- It allows for direct interaction between students and teachers. This is important for STEM education, as it allows students to get immediate feedback on their work and to ask questions.
- It allows for hands-on learning. STEM concepts are often best learned through hands-on activities, and face-to-face instruction provides opportunities for students to engage in these activities.

- It allows for collaboration. STEM problems are often best solved by working together, and face-to-face instruction allows students to collaborate on projects and assignments.

## Role and Contribution of Artificial Intelligence in STEM Teaching and Learning in Africa

According to Mutambara (2024, p. 3), "control systems" are "networks that seek to influence the behavior of a dynamic system in pursuit of specified objectives or desired responses". In this way, Artificial intelligence (AI) is a control system with interconnections of components forming configurations that facilitate the achievement of specified outcomes, in this case, the facilitation of learning. Therefore, AI as a control system can potentially revolutionize STEM education in many ways. Here are some of the key roles that AI can play in STEM teaching and learning:

***Personalized Learning:*** AI can create personalized learning experiences for each student based on their needs and interests Onesi-Ozigagun et al. (2024). This can help students learn more effectively and efficiently and stay motivated and engaged in their studies. For example, an AI system that tracks each student's progress and provides them with videos, exercises, and practice problems that are tailored to their individual needs.

***Adaptive Assessment:*** AI can provide adaptive assessment, meaning that students are given questions tailored to their individual level of understanding. This helps ensure students are challenged appropriately and receive the feedback they need to improve their learning. For example, *the MathSpring program* uses AI to provide adaptive assessment for students in grades 3–8. *The MathSpring program* analyzes each student's answers to mathematics questions and provides them with feedback tailored to their individual level of understanding.

***Virtual Labs and Simulations:*** AI can create virtual labs and simulations that allow students to experiment with STEM concepts in a safe and controlled environment. This can help students to develop their problem-solving and critical-thinking skills and to gain a deeper understanding of STEM concepts. For example, *the SimSpark platform* uses AI to create virtual labs and simulations for students in STEM subjects such as physics, chemistry, and biology. *The SimSpark platform* allows students to experiment with STEM concepts in a safe and controlled environment and to get feedback on their experiments from the AI system.

***Collaborative Learning:*** AI can facilitate collaborative learning, where students work together to solve problems and complete projects. This can help students develop teamwork and communication skills and learn from each other. For example, the Code.org program uses AI to facilitate collaborative learning for computer science students. The Code.org program uses AI to match students with partners who have similar skills and interests and to provide them with feedback on their code.

***Real-World Problem-Solving:*** AI can help students solve real-world problems using STEM concepts. This can help students see STEM's relevance to their lives and develop critical thinking and problem-solving skills. For example, *the Global Learning XPRIZE* competition uses AI to help students solve real-world problems. *The Global Learning XPRIZE* awards prizes to teams that develop the best AI-powered platform for helping students solve real-world problems.

Overall, AI has the potential to make STEM education more personalized, engaging, and relevant to students. This can help prepare students for the STEM jobs of the future and help them become more creative and innovative thinkers.

## Infusing Indigenous Knowledge Systems and Cultural, Intellectual Capital in STEM Education

It has been worrisome to note that the devaluation of indigenous African knowledge characterized pre-independence African education, the hegemony of Western forms of knowledge, and the fundamental erasure of the African people's knowledge legacy. This means that the African education system's curricula were not responsive to Africa's needs.

Using stoichiometry as an example, Bhukuvhani (2020) proposed the VALUES model (Figure 2.2) on how indigenous cultural practices interface with and influence the teaching and learning of scientific knowledge. Stoichiometry is a concept in Chemistry, one of the key STEM disciplines. It evaluates the quantification of substances in chemical reactions and processes. The VALUES model has six elements: valued knowledge, argumentation, language system, understanding, engagement, and support system.

***Valued Knowledge:*** Students tend to understand concepts related to their day-to-day practices and experiences. Such knowledge is valued as relevant in solving problems and deciding on socio-scientific issues. This reveals that science learning is not value-free. In their study in Thailand, Chonkaew et al. (2016) also report

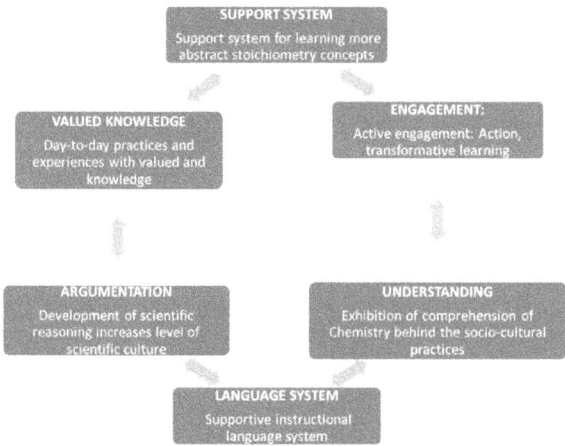

**Figure 2.2:** The VALUES model.
*Source:* Bhukuvhani, 2020, p. 217

that students hardly develop a conceptual understanding of the studied material and have no established connection to real-world issues. Students perform better on science class tasks with familiar processes or concepts. Abstract concepts with no links to real-world references for practicality and value attachment to such knowledge may be complex for students. Aikenhead (2005), in her study in Canada, also observes that students reject mechanistic mathematical descriptions of chemistry. Science problems that do not have evident value to the students' day-to-day lives are poorly performed and may not provide satisfactory explanations for the concepts contained therein.

Traditional practices and experiences have inherent, important knowledge from which the students understand the related science concepts. The traditional practices bring familiarity to the concepts related to the practices. These would have been learned experientially over a long period through the medium of their local language, which they understand well enough to express in their terms. The valued knowledge from the students' home and immediate society community of practice may increase the students' levels of scientific culture as it enhances understanding of the underlying concepts behind the practices.

The knowledge from the two thought systems is in the equipollent adaptive co-existence outcome state (Ogunniyi, 2007). Ogunniyi (2007), cited by Ogunniyi and Iwuanyanwu (2024), posits that the CAT points out that it is possible for two competing thought systems to co-exist and be equally powerful, adaptable, active, effective, or significant in making sense of the observed phenomena. Therefore, educators are proposed to design and formulate STEM discipline problems and tasks linked to carefully selected culturally-inclined practices and processes. Educators are further called upon to design problems and tasks whose knowledge is highly valued by students in their contexts. The problems, therefore, should be relevant in terms of their applicability in decision-making and solving societal socio-scientific issues. In this way, cognitive conflicts that may result from worldviews created from the two sources from which the students get their knowledge to explain phenomena, school science, and their cultural practices and experiences may be minimized. Thus, the result would promote what Sullivan (2024) terms smooth cultural border crossing in learning science concepts, such as reaction stoichiometry.

***Argumentation:*** Limited scientific reasoning may result from the rote and mechanistic learning students get from the school and their immediate social communities of practice. It may be possible that students may fail to reason logically. This may arise when they fail to explain or justify the reasons behind some phenomena, even though they might have been familiar with the practice in their past and present societal contextual practice. Toulmin's (1958) argumentation framework is relevant for developing scientific reasoning and increasing levels of scientific culture among students. Argumentation enhances critical thinking skills, one of the students' desired competencies and exit profiles (MoPSE, 2015). Toulmin's (1958) framework outlines that every knowledge claim, whether in statements about a particular phenomenon or an answer to a question, should be accompanied by adequate and appropriate evidence. The evidence may be in the form of supporting facts or statements on which the knowledge claim is based (Berland and McNeill, 2010). For example, class tasks in STEM disciplines should require students to provide explanations for their answers. Therefore, argumentation enables students to interrogate knowledge claims and test the strength of the evidence used to support or refute such claims (Zohar and Nemet, 2002). Mezirov (1997) notes that educators must consciously assist students to become autonomous thinkers by being aware and critical of their assumptions and those of others.

Equally, even in the case of rebuttals, which are also referred to as counterclaims, counterarguments (Khun, 2010) or alternative claims, Berland and McNeill (2010) say these should also be supported by providing evidence and appropriate reasoning to justify the point of view. In teaching critical thinking, even those students with different views must be given chances to present their counterarguments and the associated explanations to justify the counterclaims. In agreement with Osborne, Erduran, and Simon (2004) and Khun (1991), arguments involving rebuttals are more difficult than those without them as these require more intricate thinking. Braund et al. (2013) view argumentation as a learning method capitalizing on discussions and interaction within and among groups. The groups are collective senses and meaning-making procedures in the social construction of knowledge processes (Vygotsky, 1978) from socio-cultural practices, processes, and experiences that involve the application of scientific knowledge.

*Language System:* Language plays a significant role in learning STEM disciplines for conceptual understanding. In most cases, students have difficulties understanding concepts when English is used as the instructional language. The use of the local language helps aid them in understanding. It is not enough for educators to be conversant in either English, the official instructional language as per the policy, or the subject matter language. Rather, educators need to be competent in local languages for code-switching. Code-switching facilitates the use of local language equivalents or alternative science terminologies and explanations for phenomena to enhance students' learning in a complementary manner. Gudhlanga (2005) observes that no language is more expressive than the other.

Although African languages have been relegated to restricted code (Jones, 2013) statuses, it is important to note that if children use a language that they already know, effective education is achieved (Babaci-Wilhite, 2019). This is because a community's language is embedded with local linguistic and cultural knowledge. Nsamenang (2006) weighs in on the discourse by arguing that using students' local home languages enhances scientific reasoning, understanding of content knowledge, and self-efficacy. However, Jongore et al. (2013) lament the poor base for first languages. There have also been reports of students' poor performance, with some schools recording 0% pass rates (e.g., ZBC, 2019; Nkala, 2014; The Standard, 2014). The low pass rates have been attributed to the inappropriately deployed non-local language-speaking teachers' inability to explain the taught concepts in the students' local languages. The linguistic incompetence of the teacher results in the subject matter knowledge being inaccessible due to the language barrier.

*Understanding (**Comprehension**):* Understanding the fundamental concepts of STEM disciplines is essential for students' conceptual understanding. Students who are familiar with and value the practices involved in the class tasks (see valued knowledge above) exhibit some knowledge of the discipline behind the practices. Students who demonstrate knowledge of the STEM discipline behind traditional practices also show increased scientific literacy, thereby enhancing learning and understanding. Posner et al. (1982) explain that for a scientific concept to be accommodated for learning and understanding, it should be plausible, intelligible, and have the capacity to solve problems, as in the practicality of the socio-cultural practices in which reaction stoichiometry is utilized. Interrogating processes and critiquing answers given to explain particular phenomena also help students understand, as explained above under argumentation. An understandable language system, active engagement, and a sustainable support system are other components of this model.

***Engagement:*** The situated cognition theory views learning as participation in a community of practice. Students' engagement is a key to achieving meaningful learning. Practical learning encourages students to move from mere observers to experts by actively participating in the community of practice (Lave and Wenger, 1991). Transformative learning and the 6E+S (Engage, Explore, Explain, Elaborate, Evaluate, Extend, and Standards) instructional model possess crucial stages that involve the active engagement of students in real-world, concrete experiences and practices. The *VALUES model* is thus informed by transformative pedagogy and the 6E+S instructional model. Learning activities that engage students challenge their existing knowledge and stimulate curiosity, thereby encouraging them to pursue answers to their questions (Şahin and Kılıç, 2024).

Educators can actively engage students to tap into contextual scientific knowledge reservoirs within the local indigenous societies to promote curriculum relevancy. According to Pesanayi et al. (2019), situated engagement enhances quality education in terms of relevance. Situated engagement aids in developing higher-order systems thinking and processing skills among students. However, the solicited prior knowledge should comply with the set curriculum *standards* (the S in the 6E+S model) (Şahin and Kılıç, 2024). It can be noted that engagement is key in the application of social constructivist approaches, as the solicited prior knowledge forms the basis on which the students create their new knowledge.

Students gain knowledge through their active participation in society. Suppose such relevant cultural practices related to STEM discipline topics are identified as explained above under "valued knowledge". Some self-built experiments (Bhukuvhani et al., 2013) may be designed and carried out in that case. The self-built experimental activities make it possible for the students to realize practical learning in an unspectacular manner in rural setups and without diluting the curriculum (Bhukuvhani et al., 2012, 2013). Carefully selected and designed practical activities are important in developing conceptual understanding. Practical activities afford students authentic learning through observations, acquisition, application, and consolidation of ideas (Burke, 2009).

***Support Systems:*** Support systems that support scaffolding are necessary for STEM teaching and learning. Scaffolding students' knowledge of related concepts enhances students' conceptual understanding. These concepts may be taught first, so the students can build new knowledge on the platform set by the preceding concepts. Scaffolding follows Vygotsky's principle of the zone of proximal development. Vygotsky insists that students are helped to reach their levels of potential development by establishing first their levels of actual development. Then, students must be assisted through well-planned activities under the educator's guidance or in collaboration with other competent peers (Lui, 2012; Stewart, 2012).

Concepts become more difficult with the increase in abstraction (Mpofu et al., 2007; Gabel, 1999), cited in Thomas (2012). The learning of such abstract concepts may require support through the use of visual models, simulations, and virtual reality. The values model extends knowledge in culture-science learning discourse. Since cognitive conflicts impair conceptual understanding, Chiromo (2015, p. 164) proposes a cognitive management pedagogy cycle recognizing the need for culturally appropriate pedagogies. The VALUES model provides a possible format for culture-sensitive pedagogies to take and utilize.

## Conclusion

Africa requires reconfiguring STEM curricula to address its socio-economic and developmental challenges deliberately and adequately. Africa has implemented policies and projects to adopt STEM education as one of the strategies to address these challenges and build knowledge-based and technology-rooted economies. Some strategies include adopting transformative pedagogies such as technology-enhanced teaching and learning and culture-anchored methodologies. These are poised to enhance STEM disciplines' learning for conceptual understanding whilst transforming lives by providing solutions to real-world socio-scientific issues towards sustainable development. It is recommended that curriculum development in STEM disciplines take note of technological developments such as AI and blended learning, as well as culture, for contextual applicability and relevance, as in the proposed VALUES model.

## Bibliography

Aikenhead, G. S. (1996). Science education: Border crossing into the subculture of science. *Studies in Science Education*, 2(7), 1–52.

Aikenhead, G. S. (2005). Cultural influences on the discipline of chemistry. In *88th Canadian Chemistry Conference* (pp. 1–8). Saskatoon: University of Saskatchewan.

Aikenhead, G. S., & Jegede, O. J. (1999). Cross-Cultural Science Education : A Cognitive Explanation of a Cultural Phenomenon, 36(3), 269–287.

Al Hamad, N. M., Adewusi, O. E., Unachukwu, C. C., Osawaru, B., & Chisom, O. N. (2024). A review of the innovative approaches to STEM education. *International Journal of Science and Research Archive* 11(1), 244–252.

Berland, L. K., & McNeill, L. (2010). A learning progression for scientific argumentation: Understanding students' work and designing supportive instructional context. *Science Education* 94, 765–793.

Bhukuvhani, C. (2020). *Influence of culture on students' conceptualisation of stoichiometry at a rural secondary school in Zimbabwe.* Unpublished DPhil Thesis, Midlands State University, Zimbabwe.

Bhukuvhani, C., Chiromo, A., & Chukunda, C. (2019). Towards a framework for understanding the influence of selected aspects of culture on secondary school students' learning of stoichiometry: Insights from a preliminary study. *Zimbabwe Journal of Educational Research* 31(3), 391–417.

Bhukuvhani, C., Chiromo, A., & Chikunda, C. (2020). Influence of culture on secondary school students' learning of stoichiometry: A case of a Guruve district school, Zimbabwe. *Journal of New Vision in Educational Research* 1(2), 245–262.

Bhukuvhani, C., Mupa, M., Mhishi, M., & Dziva, D. (2012). Science practical work instructional technologies and instructional technologies in science teacher training: A case study in Zimbabwe. *International Journal of Education & Development Using Information & Communication Technology* 8(2), 17–27.

Bhukuvhani, C., Sana, A., & Tigere, A. (2013). Self-built experimental projects (SBEP): Curriculum engineering for science practical work in open and distance learning environments. *International Journal of Curriculum Development and Planning* 2(5), 1–14.

Bodner, G. M., & Bhattacharyya, G. (2005). A cultural approach to problem solving. *Educación Química* 16(2), 222–229.

Braund, M., Scholtz, Z., Sadeck, M., & Koopman, R. (2013). First steps in teaching argumentation: A South African study. *International Journal of Educational Development 33*, 175–184. <https://doi.org/http://dx.doi.org/10.1016/j.ijedudev.2012.03.007>.

Burke, K. (2009). *How to Assess Authentic Learning, 5th ed.* Thousand Oaks, CA: Corwin.

Chandrasegaran, A. L., Treagust, D. F., Waldrip, B. G., & Chandrasegaran, A. (2009). Students' dilemmas in reaction stoichiometry problem solving:

Deducing the limiting reagent in chemical reactions. *Chemistry Education Research and Practice 10*(1), 14–23.

Chiromo, A. S. (2015). *Zimbabwean Rural Day Secondary School Students' Management of Conflicts Arising from Studying the Human Reproductive System*. Unpublished PhD Thesis, University of Zimbabwe.

Chonkaew, P., Sukhummek, B., & Faikhamta, C. (2016). Development of analytical thinking ability and attitudes towards science learning of grade-11 students through science technology engineering and mathematics (STEM education) in the study of stoichiometry. *Chemistry Education Research and Practice 17*(4), 842–861.

Familoni, B. T., & Onyebuchi, N. C. (2024). Augmented and virtual reality in us education: a review: analyzing the impact, effectiveness, and future prospects of AR/VR tools in enhancing learning experiences. *International Journal of Applied Research in Social Sciences 6*(4), 642–663.

Felder, R. M., & Brent, R. (2024). *Teaching and learning STEM: A practical guide*. John Wiley & Sons.

Gadzirayi, C. T., Bongo, P. P., Ruyimbe, B., Bhukuvhani, C., & Mucheri, T. (2016). Diagnostic study on status of STEM in Zimbabwe. *Bindura University of Science Education and Higher life Foundation*.

Gauchon, L., & Méheut, M. (2007). Learning about stoichiometry: from students' preconceptions to the concept of limiting reactant. *Chemistry Education Research and Practice 8*(4), 362–375.

Gudhlanga, E. S. (2005). Promoting the use and teaching of African languages in Zimbabwe. *Zimbabwe Journal of Educational Research 17*(1), 54–68.

Jones, P. E. (2013). Bernstein's "codes" and linguistics of "deficit." *Language and Education 27*(2), 161–179.

Jongore, M., Chirimuuta, C., Bhukuvhani, C., & Zuvalinyenga, D. (2013). The Zimbabwean first languages seem to be slowly dying a natural death: A case of university students' academic writing. *Greener Journal of Educational Research 3*(7), 318–325.

Kazembe, T. C., & Musarandega, A. (2012). Student performance in A-level chemistry in Makoni district, Zimbabwe. *Eurasian Journal of Physics and Chemistry Education 4*(1), 2–29.

Khun, D. (1991). *Thinking as an Argument*. Cambridge: Cambridge University Press.

Khun, D. (2010). Teaching and learning science as an argument. *Science Education 94*, 810–824.

Lave, J., & Wenger, E. (1991). *Situated Learning: Legitimate Peripheral Participation*. Cambridge: Cambridge University Press.

Lui, A. (2012). Teaching in the zone: An introduction to working within the zone of proximal development (ZPD) to drive effective early childhood instruction. Retrieved from <www.childrensprogress.com>.

Lüy, Z., Bakirci, F., Artsın, M., Karataş, S., & Çakmak, E. K. (2024). Innovative Technologies to Trigger Creative Thinking in Micro-Lessons. In *Optimizing Education Through Micro-Lessons: Engaging and Adaptive Learning Strategies* (pp. 118–140). Pennsylvania: IGI Global.

Mekonnen, Z. B., Yehualaw, D. D., Mengistie, S. M., & Mersha, B. S. (2024). The effect of 7E learning cycle enriched with computer animations on students' conceptual understanding and overcoming misconceptions. *Journal of Pedagogical Research* 8(2), 325–356.

Mezirov, J. (1997). Transformative learning: Theory to practice. *New Directions for Adult and Continuing Education* 4, 5–15.

MoPSE. (2015). *Zimbabwe Education Blueprint 2015–022—Curriculum framework for primary and secondary Education*. Harare: Ministry of Primary and Secondary Education (MoPSE).

Mpofu, V., Kusure, L., Nhenga, J., & Zishiri, G. (2007). Advanced level chemistry students' understanding of stoichiometry: Evidence from four schools in Zimbabwe. *Southern African Journal of Education Science and Technology* 2(2), 90–100.

Mutambara, A. G. (2024). *Design and Analysis of Control Systems: Driving the Fourth Industrial Revolution*. CRC Press.

Nsamenang, A. B. (2006). Human ontogenesis: An indigenous African view on development and intelligence. *International Journal of Psychology* 41(4), 293–297.

Nite, S. B., Capraro, M. M., Capraro, R. M., & Bicer, A. (2017). Explicating the characteristics of STEM teaching and learning: A metasynthesis. *Journal of STEM Teacher Education* 52(1), 6.

Nkala, S. (2014, January 27). Gukurahundi was blamed for poor results. *Southern Eye*. Retrieved from <https://www.southerneye.co.zw/2014/01/27/gukurahundi-blamed-poor-results/>.

Ogunniyi, M. B. (2007). Teachers' stances and practical arguments regarding a science-indigenous knowledge curriculum: Part 2. *International Journal of Science Education* 29(10), 1189–1207.

Ogunniyi, M., & Iwuanyanwu, P. N. (2024). Analysis of teachers' perspectives towards the use of IKS to improve STEM education for sustainable development. *African Journal of Research in Mathematics, Science and Technology Education*, 28(3), 319-329.

Okanlawon, A. E. (2010). Teaching reaction stoichiometry: Exploring and acknowledging Nigerian chemistry teachers' pedagogical content knowledge. *Cypriot Journal of Educational Sciences 5*, 107–129.

Okere, M. I. O., Keraro, F. N., & Anditi, Z. (2012). Pupils' beliefs in cultural interpretations of "heat" associated with anger: A comparative study of ten ethnic communities in Kenya. *European Journal of Educational Research 1*(2), 143–154.

Onesi-Ozigagun, O., Ololade, Y. J., Eyo-Udo, N. L., & Ogundipe, D. O. (2024). Revolutionizing education through AI: A comprehensive review of enhancing learning experiences. *International Journal of Applied Research in Social Sciences 6*(4), 589–607.

Osborne, J., Erduran, S., & Simon, S. (2004). Enhancing the quality of argument in school science. *Journal of Research in Science Teaching 41*(10), 994–1020.

Pesanayi, T., O'Donoghue, R., & Shava, S. (2019). Think piece: Situating education for sustainable development in Southern African philosophy and contexts of socio-ecological change to enhance curriculum relevance and common good. *Southern African Journal of Environmental Education 35*, 213–221.

Posner, G. J., Strike, K. A., Hewson, P. W., & Gertzog, W. A. (1982). Accommodation of a scientific conception: Toward a theory of conceptual change. *Science Education 66*(2), 211–227.

Royal Society. (2005). Science and technology for African Development—Joint science academies' statement. *Science and Technology for African Development.* <https://royalsociety.org/-/media/Royal_Society_Content/policy/publications/2005/9647.pdf> [Accessed 14/06/2023].

RUFORUM. (2023). About STISA. <https://www.ruforum.org/Stisa> [Accessed 25/08/2023].

Şahin, Ş., & Kılıç, A. (2024). Comparison of the effectiveness of project-based 6E learning and problem-based quantum learning: Solomon four-group design. *Journal of Research in Innovative Teaching & Learning.* doi: 10.1108/jrit-09-2023-0139

Sanger, M. (2005). Evaluating students' conceptual understanding of balanced equations and stoichiometric ratios using a particulate drawing. *Journal of Chemical Education 82*(1), 131–134.

Schmidt, H.-J., & Jigneus, C. (2003). Students' strategies in solving algorithmic stoichiometry problems. *Chemistry Education: Research and Practice 4*, 305–317.

Stewart, M. (2012). Understanding learning: Theories and critique. In L. Hunt & D. Chalmers (Eds.), *University Teaching in Focus: A Learning-Centred Approach* (pp. 1–15). Victoria: ACER Press and Routledge.

Sullivan, M. (2024). The border crossed us: Enhancing indigenous international mobility rights. *Journal of Borderlands Studies* 39(2), 247–264.

The Standard. (2014, February 3). *Zimbabwe: Matabeleland's Low Pass Rate Worrying.* The Standard. Retrieved from <https://allafrica.com/stories/201402031997.html>.

Thomas, L. M. (2012). *Teaching Reactions and Stoichiometry: A Comparison of Guided Inquiry and Traditional Laboratory Activities.* Michigan State University.

Toulmin, S. (1958). *Uses of Argument.* Cambridge: Cambridge University Press.

Vygotsky, L. S. (1978). *Mind in Society.* London: Harvard University Press.

World Bank. (2023). *Advancing Post-Graduate Education in Africa Through Regional Specialization and Collaboration in Agriculture, Health, Environment, Science, Technology, Engineering and Mathematics.* Retrieved from <https://www.worldbank.org/en/events/2022/09/20/the-african-centers-of-excellence-a-pathway-towards-sustainable-development> [Accessed 31/08/2023].

ZBC. (2019). *Nkayi Schools Record a 0% Pass Rate.* Zimbabwe: Zimbabwe Broadcasting Corporation (ZBC). Retrieved from <www.zbc.co.zw>.

Zohar, A., & Nemet, F. (2002). Fostering students' knowledge and argumentation skills through dilemmas in human genetics. *Journal of Research in Science Teaching 39*, 35–62.

CHAPTER 3

# Digital Transformation in Science, Technology, Engineering, and Mathematics Education in Sub-Saharan Africa

Brantina Chirinda,[1] Eddie M. Mulenga,[1] and Gladys Sunzuma[2]
[1]University of Johannesburg
[2]Bindura University of Science Education

**ABSTRACT**
The COVID-19 pandemic caused widespread disruptions to the teaching and learning of STEM in Africa and globally. These disruptions resulted in Science, Technology, Engineering, and Mathematics (STEM) teachers facing the urgent need to migrate from traditional classrooms to remote teaching. Consequently, the COVID-19 pandemic resulted in an unprecedented transformation in African STEM education. Digital tools and resources suddenly became critical enablers in providing continuous STEM lessons to learners. Adapting STEM teaching and learning to integrate digital technologies was onerous in Africa since many teachers lacked the skills, knowledge, courage, and confidence to do so. In addition, the advancement of digitalization has evinced the existing inequalities in accessing technology in Africa. With original research findings from Namibia, South Africa, Zambia, and Zimbabwe, this chapter provides an in-depth scientific scholarship centered on STEM teaching and learning during and post the COVID-19 pandemic and the digital transformation in the four countries. This study adopted the qualitative multi-case study research design to give an in-depth description of the phenomenon under study. The Unified Theory of Acceptance and the Use of Technology model explored the factors influencing participant teachers' acceptance of digital transformation in STEM education. Our findings from the study were that the COVID-19 pandemic had presented a golden opportunity for rethinking and reimagining STEM teaching and learning in Africa despite the digital divide. The advancement of digitalization in the four Sub-Saharan African countries allows the integration of physical and digital tools and resources, which permits flexible and worthwhile STEM teaching and learning. Despite focusing mainly on secondary school STEM teachers, the reported findings in the chapter also provide affordances to other levels of education.

*Keywords:* Advancement of digitalization, COVID-19 pandemic, digital resources, online teaching and learning, UTAUT2

## Introduction

Science, Technology, Engineering and Mathematics (STEM) education in Sub-Saharan Africa faced unique challenges amidst the COVID-19 pandemic. Yet, it highlighted opportunities for innovation and resilience within the region's educational systems. STEM fields play a critical role in driving economic development, innovation, and social progress, making their continued advancement crucial, especially during times of crisis.

When the pandemic hit, schools across Sub-Saharan Africa, like those worldwide, were forced to close their doors temporarily to curb the spread of the virus.

The closure of educational institutions led to challenges that comprised interrupted learning, higher drop-out rates, and lowered academic achievement grades (Muck et al., 2021). The closure of educational institutions moved numerous education systems worldwide to adopt remote teaching and learning (Mukuka et al., 2021), and sub-Saharan African countries were no exception. The typical face-to-face education in conventional classroom settings had to be shifted to remote teaching using technology (Mardini et al., 2022). This abrupt shift to remote learning posed significant challenges for STEM education, as access to technology and the internet is not evenly distributed throughout the Sub-Saharan African region. Many students lacked the necessary devices and reliable internet connections, exacerbating existing inequalities in educational opportunities (Chirinda et al., 2021).

Nonetheless, such massive alterations brought opportunities and challenges to the teaching and learning of STEM in both teacher education and schools. There is the possibility that STEM education will primarily be online in the near future (Ogbonnaya et al., 2020), hence the need to have an insight into STEM teaching and learning during and post the COVID-19 pandemic. The current study might guide future studies to understand the problem and possible struggles of the sudden shift to online learning and possibly accommodate future challenges faced during STEM teaching and learning in similar outbreaks.

Even though the whole education system was affected during the COVID-19 pandemic, this study has afforded special consideration to STEM education in Namibia, South Africa, Zambia, and Zimbabwe. This was necessitated by the deficient performance in STEM subjects in sub-Saharan African countries even before the COVID-19 outbreak (Bethel, 2016). Regarding such a problem, questions were raised concerning teaching practices that could be beneficial during and after the COVID-19 pandemic. Hence, this study explores digitalization in STEM teaching and learning in Sub-Saharan Africa during and after the COVID-19 pandemic. The following research question was formulated: What is the state of digital transformation in STEM education in Sub-Saharan Africa?

## Emergency Remote Teaching and Learning

An emergency remote teaching and learning approach is a form of online learning which is a temporary measure that enables continuing education in times of crisis (Hodges et al., 2020). In search of solutions to the COVID-19 crisis, the education system moved to a fully remote teaching and learning environment with various virtual communication software (Microsoft Teams, ZOOM, Google Classroom) (Chirinda et al., 2022). Emergency remote teaching and learning ensures that teachers and students are connected and engaged with the content while working from their comfort zone (Mukuka et al., 2021). The transition to remote teaching and learning using technology is flexible, allowing students to join classes from their comfort places and, in some instances, to watch the classes several times at their own pace, as well as reducing the dangers in the traveling required to attend face-to-face classes (Mardini et al., 2022). Scholars (O'Connell et al., 2022) have observed that online learning, as emergency remote teaching, offers opportunities to gain professional and technical skills through students' self-development skills. Students become more self-independent, increasing their responsiveness regarding self-development abilities to create their learning environment to attain professional skills. As a result, their professional decision-making and problem-solving skills are enhanced (O'Connell et al., 2022). However, the transition to emergency remote teaching requires technological tools and skills, internet access, and communication and lecturing software (Mardini et al., 2022).

Internet challenges, for example, weak signals or slow internet speed, may lead to technical problems affecting teachers and students (Mardini et al., 2022). Additionally, emergency remote teaching and learning require technical skills and knowledge for teachers and students to enable them to navigate communication software such as ZOOM and Google Classroom and solve any technical problems arising during live sessions (O'Connell et al., 2022). Most STEM teachers and students in Sub-Saharan Africa were forced to adapt or learn technological skills and knowledge they had not engaged with before the COVID-19 pandemic.

The abrupt transition to remote teaching and learning in sub-Saharan African countries had devastating effects because such countries have been struggling to provide quality education due to economic hardships. The COVID-19 pandemic added more challenges to the already suffering education system (Chirinda et al., 2021). In most countries, STEM teaching is done in a traditional, face-to-face way with limited use of technological tools. The pandemic forced most institutions to migrate to remote teaching without adequate planning and resources. As a result, some teachers and students were disadvantaged, whilst some continued their lessons as usual. Some institutions' unavailability of technological tools led to the digital divide and unequal STEM teaching and learning access during the COVID-19 pandemic. Access to education during the pandemic was determined by the type of school and the nature of students and teachers. Teachers and students with access to technological tools, knowledge, and skills benefited more than those from disadvantaged schools and families (Chirinda et al., 2021).

The forced remote teaching and learning that the teachers and students were engaged in during the COVID-19 pandemic might not resemble what traditional online education should be, and such an abrupt transition to remote learning might have tarnished the status of online education (Kim, 2020). Quality online learning programs require high-input operations, development time, and significant investments (Kim, 2020). However, Kim (2020) did not imply that the COVID-19-imposed universal transition to remote teaching was bad for student learning. The need for teaching and learning with asynchronous (Canvas, Blackboard, D2L) and synchronous (Microsoft Teams) platforms might produce substantial benefits when these approaches are layered into face-to-face instruction after the pandemic (Kim, 2020). Teachers and students had a shared understanding that technological tools are complements, not substitutes, for the intimacy and proximity of face-to-face learning after the pandemic. STEM courses may be taught by moving content online, whilst valuable classroom time can be productively used for discussion, problem-solving, and guided practice.

As the world progresses towards technology-aided learning after the pandemic, a rise in the use of technological platforms and modes of instruction is anticipated. In addition, more educational institutions will switch to online mode for assessment and evaluation, and students may be further inclined to take various short-term online courses. The post-pandemic will change how educational institutions plan for, manage, and fund online education. The blended learning approach is the way to go with the Fourth Industrial Revolution.

## Theoretical Framework

The intention of individuals to use technological innovations is explained by numerous theoretical models derived from information systems, psychology, and sociology. Various frameworks have emerged to explore how individuals adopt and utilize technology. These frameworks encompass the likes of Rogers' (1995) Theory of Diffusion of Innovations, Davis et al.'s (1989) Technology Acceptance Model, Venkatesh et al.'s (2003) Unified Theory of Acceptance and Use of Technology, and the second Unified Theory of Acceptance and Use of Technology framework (Venkatesh et al., 2012). Mishra and Koehler (2006) established the Technological Pedagogical Content Knowledge framework to address the absence of a guiding theory for successfully incorporating technology in education. This study adopted Venkatesh et al.'s (2012) Unified Theory of Acceptance and Use of Technology (UTAUT2) framework in Figure 3.1. The UTAUT2 framework was appropriate for the study because we focused on digitalization in STEM education in Sub-Saharan Africa. We could achieve the utmost explanatory power by utilizing this model compared to all other standard acceptance models. This, in turn, can significantly aid the technology development process in Sub-Saharan Africa. Several studies have utilized the UTAUT2 model to explore the acceptance and utilization of technology in education. For instance, Raman and Don (2013) employed the UTAUT2 model to investigate the factors influencing pre-service teachers' adoption of a learning management system.

The UTAUT framework was developed when Venkatesh et al. (2003) identified the limitations of the existing theories. One significant limitation was the absence of empirical testing and comparison among the prevalent technology acceptance models, leading to uncertainty about the predictive ability of the constructs in each theory. Moreover, research on technology usage behavior had predominantly centered on simpler systems, with little attention given to more intricate technologies (Venkatesh et al., 2003). Venkatesh et al. (2003) proposed the UTAUT by synthesizing the propositions of the existing technology acceptance models. The UTAUT theoretical model proposes that people's use of technology depends on their intention to use it. Four key factors influence the likelihood of adopting a technology: performance expectancy, effort expectancy, social influence, and facilitating conditions (Yunus et al., 2021).

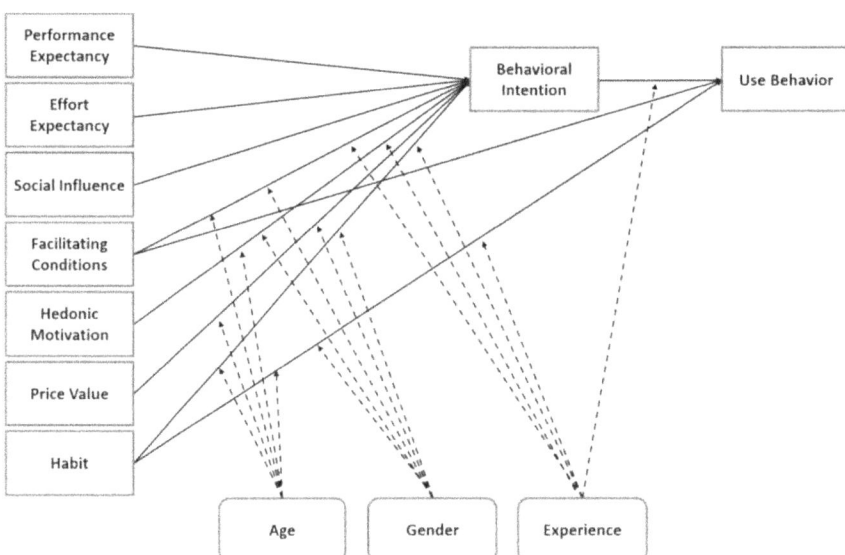

**Figure 3.1:** Unified theory of acceptance and use of technology (UTAUT2) framework.

According to Venkatesh et al. (2003), performance expectancy refers to an individual's belief that using a system will improve their job performance. Effort expectancy refers to the ease of using a system (Venkatesh et al., 2003). It is derived from various factors such as perceived ease of use, complexity, and similar definitions and scales from models such as the Technology Acceptance Model and Theory of Diffusion of Innovations. Social Influence refers to an individual's perception of important others' beliefs regarding using a new system (Venkatesh et al., 2003). Venkatesh et al. (2003) noted that the impact of social influence is significant when technology use is mandatory. Individuals may use technology due to compliance requirements, not personal preferences (Venkatesh and Davis, 2000). Facilitating conditions refer to the extent to which an individual believes that an organization's technical infrastructure is in place to support the use of a system (Venkatesh et al., 2003). The facilitating conditions construct was formed from various constructs drawn from different models. Compatibility, perceived behavioral control, and facilitating conditions are among these constructs. Facilitating conditions directly impact the intention to use, but the effect becomes insignificant after initial use. The first three constructs directly determine the intention to use new technology. Facilitating conditions determine the intention to use new technology and subsequent user behavior.

Through empirical analysis, the UTAUT model explores various technology acceptance theories. While these theories may provide different or incomplete perspectives, UTAUT boasts a more substantial predictive power by explaining 70% of the variance in use intention (Venkatesh et al., 2003). In contrast, other models examining technology acceptance, such as Davis (1993), account for less variance. UTAUT further highlights that personal and demographic factors, like age, gender, and experience, can influence technology acceptance by interacting with certain constructs (Venkatesh et al., 2003). As computing technologies such as gaming and social media advanced, the original UTAUT model failed to explain why users voluntarily engage with such platforms. As a result, Venkatesh et al. (2012) added three new constructs to the original four key determinants of user acceptance and use of technology. This led to development of the second version of the UTAUT2. These new constructs are hedonic motivation, price value, and habit. The three constructs attempt to answer the questions: Is using the internet enjoyable and cost-effective for consumers? Do consumers have established habits for internet use?

## Methodology
### Research method
This study adopted a multiple-case qualitative research design to give an in-depth description of the phenomenon under study (Yin, 2017). Yin suggests that case studies help explain, describe, or explore events or phenomena in their everyday contexts. For instance, in our research, they helped us understand digital transformation in STEM education in Sub-Saharan Africa. With original research

findings from Zambia, South Africa, Namibia, and Zimbabwe, this chapter provides an in-depth scientific scholarship centered on STEM teaching during and after the four countries' COVID-19 pandemic and the digital transformation.

## Participants

In total, 16 (10 females and six males) teachers were purposively selected to participate in the study—five from Zambia, four from South Africa, three from Namibia, and four from Zimbabwe by responding to a questionnaire designed to investigate STEM teaching during and post the COVID-19 pandemic and the digital transformation in the four countries. Participants between 25 and 65 years old were selected because they taught at public schools during and after the COVID-19 pandemic. The in-service teaching years of the participants ranged between 3 and 35 year. Participation was voluntary, and participants were assured of anonymity as no personal information was collected for confidentiality purposes.

## Research instrument

To address the research question(s), we designed a research questionnaire that comprised three sections. We grouped the questionnaire items into three sections. The first section was dedicated to *Demographic Details* of teachers (i.e., gender, age, country, and city). The second section was devoted to *Mathematics Teacher Professional Information*, particularly emphasizing participant teachers' educational and professional qualifications, total years of teaching, and STEM subjects currently being taught by participants. The last section dealt with STEM teaching during and post the COVID-19 pandemic and the digital transformation in the four countries. The research questionnaire underwent a series of validation processes before finally being implemented in the purposively selected sample. First, after the questionnaire was structured, it was shared among the three researchers for face validation, clarification of items, highlighting some of the key factors relevant to the study, and verifying if it measured the intended variables. Positive comments and suggestions were recorded, and the questionnaire was modified. Second, a pilot test of the questionnaire was conducted by the three authors with a sample of four mathematics teachers, each coming from the four countries; based on the participants' responses, minor modifications concerning the wording of some statements were produced, which helped to assess the construct validity of the items measured in the questionnaire. The pilot study also assisted the authors in identifying potential problems that could interfere with their results transferability, confirmability, dependability, and credibility. Eventually, a refined questionnaire with a few additions emerged from the pilot testing that was finally implemented for 16 study participants. This lengthy validation procedure was undertaken to make the research instrument reliable for data collection and to ensure that the respondents did not misunderstand all the items in the questionnaire. Despite the different focus of the items in each section, all items were focused on STEM teaching in the context of the pandemic.

### Data analysis

Data were qualitatively analyzed using a standard qualitative content analysis technique. Qualitative content analysis involves systematically coding and identifying themes in text data for subjective interpretation (Hsieh and Shannon, 2005). Qualitative content analysis was relevant for this study because it is a research method that goes beyond simply counting words or extracting objective content from texts (Zhang and Wildermuth, 2009). It involves analyzing the meanings, themes, and patterns that may be explicit or implicit in a text. This method allows researchers to understand social reality in a subjective yet scientific manner. The teachers' responses were coded through multiple rounds of analysis ranging from—connecting themes, establishing patterns in the data, making notes, listing emerging codes, constant comparisons and multiple reading, and re-reading stages (Creswell, 2013).

Researchers must create transparent coding processes and conclude the data for qualitative content analysis to be credible. Coders need precise definitions and transparent procedures to ensure accurate coding and interpretation. Since three authors were coding the data, we created a manual to ensure consistent and correct coding. The coding manual included category names, code assignment rules, and examples (Weber, 1990). The coding manuals had an extra section intended for taking notes while coding. These notes were shared among coders and used to update and improve the coding manual with interpretive insights.

The identified themes were cross-checked against the categories of usage behaviors reflected in the UTAUT2 model. This was done to verify whether the observed usage patterns among teachers reflected the seven key determinants: performance expectancy, effort expectancy, social influence, facilitating conditions, hedonic motivation, price value, and habit (Venkatesh et al., 2012).

## Findings and Discussion

This study aimed to investigate digital transformation in STEM education in Sub-Saharan Africa. This section discusses teachers' acceptance and use of technology in their teaching practices.

## Performance expectancy

Performance expectancy refers to the work teachers do every day in STEM education. During our data analysis, we explored whether STEM teachers were influenced by the advantages of using technology in their work. We found that online learning platforms and learning management systems have expanded in South Africa, Namibia, and Zimbabwe. For instance, Teacher A from Zimbabwe said:

> Various digital tools, platforms, televisions, and radios were used during and even after the COVID-19 pandemic to ensure STEM teaching and learning.

This provides students and teachers access to various STEM resources, including video lectures, interactive simulations, and assignments, which align with earlier findings by Haleem et al. (2022). Teachers can now continuously deliver STEM lessons to students, even in some remote settings, which was not feasible before the digital transformation. This shift presents an excellent opportunity to rethink and reimagine STEM education in Africa, and teachers are now more enthusiastic about incorporating digital technologies into their teaching practices.

The research findings from Zambia revealed that the teaching practices of all five participant in-service mathematics teachers—before, during, and after—the COVID-19 pandemic have been face-to-face. It is of considerable interest to note that in Zambia, the teaching methods and the conditions of teaching STEM in secondary schools have never changed before, during and after the pandemic crisis, despite witnessing the devastating effects of coronavirus worldwide. In addition, one of the questions the researcher(s) sought to find answers to is cited below:

> If you are using online or blended instruction, what modes of delivery are you using?

Even when Zambia was overwhelmed by the effects of the pandemic in terms of high rate of infection transmission between teachers–and–pupils, increasing death rates among teachers and low numbers of recoveries, surprisingly, in responding to the above question, the participants in this study still narrated that they were not using online or blended instruction as a mode of delivery (see Table 3.1). This is probably because of the digital divide in Zambian secondary schools. The qualitative analysis of teachers' responses to the learning and teaching arrangements at their schools during the COVID-19 pandemic highlighted that contact hours were reduced, classes were divided to observe a 1-m apart rule, two streams were introduced to combat the spread of COVID-19 virus between teachers and mathematics and adhere to other related health guidelines as summarized in Table 3.1, supported by excerpts from the Zambian teachers' responses.

Table 3.1: Zambian teachers' responses to mathematics teaching during the pandemic era

| Mathematics Instruction During the COVID-19 Pandemic | Excerpts from Teachers' Responses |
| --- | --- |
| Mathematics teaching practices before the COVID-19 pandemic | "Face – to – face." |
| Mathematics content delivery to learners during the COVID-19 pandemic | "Face – to – face." |
| Modes of delivery | "Not applicable (N/A)." |
| Teaching and learning arrangements at school | "The arrangement is such that the number of pupils per grade has been reduced from 10 to 6. Classes which had more than 35 pupils have been split in two. The sitting arrangement according to the COVID-19 guidelines is 1 meter apart." (Zambia, secondary mathematics teacher) |
| | "The teaching period has been reduced from 6 periods per week for each class to 4 periods per week. Number of minutes per period is still 40 minutes. Each learner sits alone on the desk and there is a meter distance between the desks. Teacher pupil ratio is 1:40." (Zambia, secondary mathematics teacher) |
| | "There are two streams, there are those that come in the morning. That is, Grade 8s, 9s and 12s and those that come in the afternoon the Grade 10s and 11s. COVID-19 guidelines are strictly followed and observed." (Zambia, Mathematics teacher) |
| | "Teaching hours – 4 hours per session – and we have two sessions. Sitting arrangement: 1 meter apart and 24 [learners] per class." (Zambia, mathematics teacher) |
| How often do you see your learners for lessons? | "Contact hours, 1 hour 20 minutes per day, two times per week." (Zambia, mathematics teacher) |
| | "The contact hours with my learners per week is 4 hours." (Zambia, mathematics teacher) |
| | "Learners are seen for a period of 160 minutes weekly." (Zambia, mathematics teacher) |
| | "I meet each class twice per week. Number of minutes is 80 for two periods." (Zambia, mathematics teacher) |
| | "I have the morning session which runs from 07:30 hours to 11:50 hours and the afternoon session which runs from 12:10 hours to 16:30 hours." (Zambia, mathematics teacher) |

## Effort expectancy

Our findings revealed that while many institutions launched STEM education initiatives across Sub-Saharan Africa in response to the COVID-19 pandemic, some STEM teachers have been reluctant to embrace various e-learning platforms and have found adapting to new technologies and teaching methods challenging. This has led to some teachers feeling overwhelmed and discouraged, and as a result, they have been less inclined to use the available online tools. One Zimbabwean Teacher, S said:

> Transitioning everything to an online format can indeed be overwhelming, especially considering the time and effort required to ensure the content is interactive and engaging for students. This process often involves rethinking lesson plans, adapting materials, and familiarising oneself with new technologies and online platforms. Additionally, finding or creating interactive activities that effectively engage students in an online environment can be time-consuming.

Teachers from Namibia and South Africa highlighted similar findings. Teacher M from Namibia articulated that:

> The demands of managing online instruction left teachers with limited time for tasks such as grading and providing feedback on student work. The shift to online teaching often requires instructors to juggle multiple responsibilities simultaneously, including troubleshooting technical issues, facilitating online discussions, and supporting students in navigating the challenges of remote learning.

Such findings align with (DeCoito and Estaiteyeh, 2022), who pointed out that teachers were overwhelmed with online activities during the pandemic. Hodges et al. (2020) argue that effective online learning hinges on several critical factors, including meticulous instructional design, thorough planning, and robust development of educational materials, coupled with adequate investment in support systems. However, they suggest that these conditions may be compromised during emergencies such as the COVID-19 pandemic, potentially diminishing the quality of online teaching. In emergency scenarios like the COVID-19 pandemic, educational institutions are often compelled to transition rapidly to online learning without the luxury of sufficient time for comprehensive instructional design and planning. As a result, educators may need to hastily repurpose existing materials or develop new content under time constraints, potentially compromising the quality and effectiveness of online instruction.

The STEM teachers in the four Sub-Saharan African countries found it easier to teach using the WhatsApp platform during and after the COVID-19 pandemic than other platforms because it was already installed on their mobile phones and accessible through mobile phone service providers. Teacher P from Namibia stated that:

> WhatsApp is highly effective as an educational tool, particularly in reaching learners quickly and facilitating communication and collaboration. It is simple to use, and learners and teachers are familiar with the app.

The simplicity and familiarity of the platform have made it an attractive option for teachers and students, particularly in remote or underprivileged areas, as reported by Nhongo and Tshotsho (2021). However, it is essential to note that these feelings of reluctance are not unique to STEM teachers in Sub-Saharan Africa. The sudden transition to remote learning has been challenging for teachers worldwide. It has required them to learn new skills, such as video conferencing software and online collaboration tools, and rethink how they deliver instruction. Despite these challenges, STEM teachers in Sub-Saharan Africa must overcome their reluctance to adopt new technologies and teaching methods. By embracing online platforms and tools, teachers can provide students with a quality education, even in remote or underprivileged areas.

Fortunately, post-COVID-19, many organizations offer teachers training and support in delivering STEM education in a virtual or hybrid environment. These resources can help teachers feel more confident and comfortable with the new technologies and teaching methods. Additionally, many STEM initiatives are developing innovative and engaging online resources to help teachers deliver instruction more effectively. Nevertheless, while some STEM teachers in Sub-Saharan Africa have been reluctant to embrace remote learning platforms, they must overcome their hesitations and adapt to the new technologies and teaching methods. Despite the benefits of using WhatsApp for remote STEM education, it is important to note that the platform has limitations. It is unsuitable for delivering live lectures or interactive lessons and may not be the best option for larger class sizes. However, it can effectively supplement other remote learning platforms, especially for small group discussions and one-on-one interactions. The COVID-19 pandemic has highlighted the need for digital literacy and technology in education, and STEM fields, in particular, require a strong foundation in these areas. By embracing online platforms and tools, teachers can provide students with a quality education, even in remote or underprivileged areas.

### Social influence

Regarding incorporating technology into their teaching practices due to social influence, many STEM teachers in Sub-Saharan Africa had not received training pre- and during the COVID pandemic and were receiving limited post-pandemic training. As a result, they may have felt overwhelmed or intimidated by the prospect of navigating complex digital tools, leading to a reluctance to integrate them into their pedagogy. In Zimbabwe, the Ministry of Primary and Secondary Education (MOPSE), the government, and organizations such as UNESCO and UNICEF have influenced the use of digital tools in STEM teaching and learning.

They partnered to find innovative ways for the continued learning and teaching of STEM in Zimbabwe. For example, Teacher F from Zimbabwe said:

> MOPSE, through the government, broadcasts live lessons on *televisions* and radios, making it possible for learners to learn during the lockdowns. This influenced both the teachers and learners, and we ended up thinking of ways to make learning successful during the lockdown.

Teacher A from Zimbabwe said:

> "Dzidzo Paden Imfundwe'ndlini," a WhatsApp platform designed for remote teaching to facilitate continued learning for learners *during* periods of home isolation, was introduced by UNESCO. This digital platform played a crucial role in ensuring that STEM education could continue despite disruptions caused by factors such as the COVID-19 pandemic.

By leveraging WhatsApp, a widely accessible messaging platform, UNESCO provided an avenue for educators to deliver educational content directly to students' smartphones, enabling remote learning from the safety of their homes. This initiative helped mitigate the impact of school closures and other restrictions on traditional classroom instruction by offering an alternative means of accessing educational resources. Using digital platforms like Dzidzo Paden Imfundwe'ndlini underscores the importance of innovative approaches to education delivery, particularly in times of crisis or emergency (Thabela et al., 2020). By harnessing technology to reach students remotely, UNESCO demonstrated a commitment to ensuring that learning opportunities remain accessible and equitable, regardless of external challenges or limitations.

## Facilitating conditions

Facilitating conditions refer to an individual's perception of the resources and support available when using a particular technology product. In the case of the teachers who participated in the study, this perception was closely linked to easy access to technology and the availability of technical training and support from their school and government. For instance, Teacher S from Zimbabwe said:

> The continued STEM learning and teaching were facilitated by the radio lessons programme broadcasted on National FM, Power FM, Classic 263, Radio Zimbabwe, and Khulumani FM, initiated by the government and MOPSE. The radio lessons mitigated internet challenges.

Community radio and television broadcasts are emerging as practical tools for reaching students, particularly in regions with limited internet access or connectivity challenges (Ayanwale et al., 2023). Research suggests that radio broadcasts may be particularly effective in reaching students during the pandemic and beyond (Ayanwale et al., 2023). The accessibility and widespread availability of radio make it a powerful medium for delivering educational content to diverse audiences, including those in remote or underserved areas. Additionally, efforts to leverage television and radio, individually or in combination, have been most active in Africa, highlighting the region's commitment to finding creative solutions for learning during crises (Ayanwale et al., 2023). Rural teachers also mentioned the lack of adequate infrastructure, such as broadband internet access and reliable electricity. Urban areas tend to have better infrastructure than rural areas, leading to a significant gap in access to digital technologies. The digital divide between rural and urban schools in Sub-Saharan Africa is a considerable challenge that has hindered the progress of STEM education on the continent.

Our findings revealed that while urban schools enjoy the benefits of technology, such as access to computers, high-speed internet, and other digital resources, rural schools often need more such resources. This disparity in resource allocation has created a gap between the quality of STEM education in urban and rural areas, making it difficult for rural students to compete with their urban counterparts before, during, and after the COVID-19 pandemic. Rural schools' lack of access to technology directly impacts digitalization in STEM education in Sub-Saharan Africa. For instance, without computers and internet access, students in rural schools cannot access digital resources that can enhance their STEM learning experience (Moonasamy and Naidoo, 2022). Rural students may be disadvantaged in knowledge acquisition, digital literacy, and practical skills required in STEM education. As a result, the digital divide between rural and urban schools in Africa has widened the gap between those with access to quality STEM education and those without (Moonasamy and Naidoo, 2022). Our findings revealed that one of the main reasons for the digital divide in Africa is the lack of infrastructure in rural areas. Most rural areas lack basic infrastructure, such as electricity and internet connectivity, essential for integrating technology into STEM education (Nhongo and Tshotsho, 2021). Additionally, a shortage of qualified teachers and technical support staff in Sub-Saharan African rural schools make implementing technology in STEM education difficult.

### Hedonic motivation

In the context of UTAUT2, hedonic motivation is one of the key determinants influencing users' behavioral intentions and actual usage of technology. According to the framework, individuals are more likely to adopt and use a technology if they perceive it as enjoyable and satisfying, regardless of its instrumental benefits

(Venkatesh et al., 2012). Participant teachers preferred WhatsApp to dedicated educational platforms like Canvas, BlackBoard, bCourses, etc. It is possible that the preference for WhatsApp over official educational platforms may be due to the enjoyment users derive from using WhatsApp. Some teachers argued that their students felt more comfortable contacting them or their classmates on WhatsApp for quick questions, clarifications, or discussions that may not warrant formal communication through dedicated educational platforms. Teachers did not feel it was important to invest their time in becoming familiar with a dedicated educational platform. This was articulated by Teacher D from South Africa:

> Most of us enjoy using WhatsApp to communicate with friends and family. Integrating academic communication into a platform we use regularly enhances engagement and participation, so there is no need to use the blackboard.

Zimbabwean teachers expressed similar sentiments about the enjoyment of WhatsApp use. Teachers' preference for WhatsApp can be explained by the fact that once individuals develop a habit of using a particular technology and derive pleasure from it, they are more likely to continue using it regularly, even without explicit utilitarian benefits. While WhatsApp offers several advantages for informal communication and collaboration, it is essential to recognize that the educational platform serves a different purpose as a comprehensive learning management system designed to facilitate structured online learning experiences, including content delivery, assignments, grading, and course management (Mulyono et al., 2021). In some cases, both platforms can complement each other, with WhatsApp as a supplementary communication tool alongside a dedicated educational platform to enhance engagement and interaction among students and teachers.

### Price value

Price value refers to how users can be influenced to adopt a particular technology based on the monetary cost or who bears the cost. We investigated whether any outcomes were associated with the technology options available to the STEM teachers. Our findings were that the COVID-19 pandemic highlighted the importance of digital resources in STEM education. With the shift to remote learning during the COVID-19 pandemic, students and teachers relied heavily on digital resources to continue their education. However, the high cost of data has made it difficult for many to access these resources, leading to further inequalities in STEM education (Moonasamy and Naidoo, 2022). This was indicated by Teacher C from Namibia:

> STEM education is increasingly important today, providing students with the skills and knowledge *needed* to succeed in a rapidly changing job market. However, Namibia's high data cost has limited access to digital resources essential in STEM education.

Our findings revealed that the high cost of data in Sub-Saharan Africa significantly impacts digital transformation in STEM education in Africa. Teacher B from South Africa stated that:

> STEM education requires access to digital resources such as online textbooks, videos, and virtual labs. However, the high cost of data in South Africa has made it difficult for many students, particularly those living in rural or low-income areas, to access these resources.

Digital resources are essential in providing students with a hands-on and interactive learning experience. The cost of data is a significant challenge for STEM educators as well. Teachers often need to access digital resources to prepare lessons and assignments, but the high cost of data limits their ability to do so (Moonasamy and Naidoo, 2022). On another note, teachers highlighted using televisions and radios to promote the teaching and learning of STEM subjects during the COVID-19 era. A Zimbabwean Teacher, F, had this to say:

> Televisions and radios were being used to deliver lessons on STEM subjects. With the present ICT infrastructure in Zimbabwe, radio is the best mode of delivering lessons, supporting the continuity of learning. However, there are costs associated with using televisions and radios, such as electricity charges. Learners in some rural settings were still facing some challenges as there were no television and radio signals in their geographic locations.

Some teachers from South Africa also highlighted similar sentiments about lessons being delivered through television and radio. The findings align with those of Nhongo and Tshotsho (2021), who highlighted the signal challenges in some areas in Zimbabwe. Our findings revealed that internet-based platforms such as Google Classroom, Zoom, Microsoft Teams, and Google Meet were used in non-government Zimbabwean schools and presented opportunities and challenges for educational access and equity. Zimbabwean teacher S stated that:

> Internet-based platforms such as Google Classroom, Zoom, Microsoft Teams and Google Meet were used in non-government schools. However, access to such platforms was provided for the learners who would have paid fees. Therefore, not all learners from non-government schools had access to such platforms because learning was based upon payment for the service.

Nkomo and Matli (2022) observe that while these platforms offer powerful tools for remote teaching and learning, access to them may be restricted for students who cannot afford the associated fees. In non-government schools, where access to internet-based platforms is contingent upon payment of fees, students from lower-income households may be disproportionately affected. This creates a digital divide, where students with financial means can access online learning resources. At the same time, those from economically disadvantaged backgrounds may be excluded from participating fully in remote STEM education (Moonasamy and Naidoo, 2022). The reliance on fee-based models for access to internet-based platforms exacerbates educational inequalities and perpetuates disparities in learning opportunities.

Governments and policymakers need to take urgent action to address the high data costs in STEM education. One solution could be to provide subsidies for data costs for students and teachers. Additionally, investing in infrastructure development could reduce the overall cost of providing internet services, making it more affordable for everyone.

In conclusion, the high cost of data in Sub-Saharan Africa is a significant challenge for STEM education. Access to digital resources is essential in providing students with the skills and knowledge needed to succeed in today's rapidly changing job market. Addressing this challenge requires the collaborative efforts of governments, policymakers, educators, and other stakeholders to ensure that everyone has access to affordable data. It is important to note that digital textbooks are becoming more prevalent post-COVID pandemic, often tailored to the African curriculum. They are cost-effective and easily accessible. The Open Educational Resources initiatives provide free and open access to a wealth of STEM educational materials, including textbooks, videos, and interactive modules, benefiting both students and educators.

## Habit

The UTAUT2 model suggests that individuals who have developed a habit of using a particular technology are likelier to continue using it regularly, even without explicit benefits (Venkatesh et al., 2012). The teachers were proficient in using the WhatsApp interface for daily communication. As a result, Sub-Saharan African teachers naturally used the platform to engage with their students and continue their teaching activities during the COVID-19 pandemic. In addition, our findings were that WhatsApp became a popular choice for teachers due to its ease of use, accessibility, and wide range of features, allowing effective communication and collaboration. Teymori and Fardin (2020) highlighted that many students and teachers may need help using new technologies and digital tools required for remote learning. This lack of familiarity can hinder the effective implementation of online instruction and create barriers to accessing educational resources. In this study, most teachers resisted adopting other platforms like Canvas, BlackBoard, etc., due to the additional challenges that learning new interfaces would bring them.

Teacher V from South Africa commented:

> One of the main advantages of using WhatsApp is that it allows me to communicate with my students in real-time. This is especially important during remote learning, as it enables me to provide immediate feedback, answer questions, and address concerns

as they arise. However, I am not comfortable learning to use new platforms. I believe it isn't easy and will take my time.

Some teachers felt no need to use other platforms because WhatsApp is free and requires no special equipment or software. This makes it accessible to teachers and students in low-income areas, who may not have access to expensive technology or high-speed internet. As noted by Bouhnik and Deshen (2014), this aspect of WhatsApp's accessibility contributes to its widespread adoption and use, as it lowers barriers to entry for individuals who may not have the means to purchase expensive smartphones or pay for subscription-based messaging services. Furthermore, WhatsApp is available on a wide range of devices, including smartphones, tablets, and desktop computers, which means that students can access the platform from anywhere and at any time. The teachers used WhatsApp, which is supported by Nokia, Blackberry, iPhones, and Android smartphones, and it operates on various current devices and operating systems (Sunzuma et al., 2022). Teachers reported that WhatsApp facilitated teaching as they could utilize multiple platform features to enhance the learning experience. Specifically, they could use audio recordings to teach selected mathematics topics, save the text of discussions, and share images and videos with their learners. Teacher E articulated this from Zimbabwe:

> I see no need to use other platforms because WhatsApp is free and allows for group chats, which allow for more interactive and engaging student discussions like those on the Blackboard and Canvas platforms. I can also share multimedia resources, such as videos, images, and audio files, which help reinforce learning and make the lessons more engaging.

The findings indicate that teachers utilized several functions of WhatsApp effectively to deliver online lessons during the COVID-19 period. WhatsApp's various functions facilitated easy accessibility of learning resources, aligning with earlier research by Bonsu et al. (2021). The findings align with the earlier findings by Sunzuma et al. (2022), who pointed out such beneficial aspects of WhatsApp. Our analysis indicated that WhatsApp has been a helpful tool for many STEM teachers during and after the COVID-19 pandemic; it is important to note that its use may not be suitable for all teaching activities (Sunzuma et al., 2022). For example, it may not be appropriate for sharing sensitive information or conducting assessments.

Additionally, some students may not have access to smartphones or other devices required to use WhatsApp, which can create inequities in learning opportunities (Nhongo and Tshotsho, 2021). In conclusion, WhatsApp has emerged as a popular tool for teachers during and after the COVID-19 pandemic due to its accessibility, ease of use, and ability to facilitate real-time communication and collaboration. While it may not be suitable for all teaching activities, it has helped bridge the gap between teachers and students during a challenging period of remote learning at the peak of the COVID-19 pandemic.

## Conclusion

Digital transformation in STEM education is crucial for the development and growth of the Sub-Saharan African region. The integration of technology in STEM education can improve student engagement and learning outcomes, as well as enhance teaching and learning processes. However, some challenges, such as the digital divide and access to technology, security risks, and data privacy concerns, must be addressed. Addressing the digital divide is critical. Ensuring students have access to devices and reliable internet connections remains challenging, especially in rural and underserved schools. Both students and teachers need training in digital literacy to use digital tools effectively for STEM education.

Despite these challenges, the benefits of digital transformation in STEM education in Sub-Saharan Africa are significant. It can improve student performance and collaboration, and communication between students and teachers. It has the potential to enhance teaching and learning processes. Furthermore, it can give students the necessary skills to succeed in the digital age. Collaboration between stakeholders is essential to implement digital transformation in STEM education in Sub-Saharan Africa. This includes government agencies, educational institutions, and the private sector. Additionally, professional development and training for teachers can help them to use technology in the classroom effectively. In conclusion, digital transformation in STEM education is crucial for developing Sub-Saharan Africa.

## Bibliography

Ayanwale, M. A., Adewuyi, H. O., & Afolabi, O. W. (2023). Learning through radio and television during COVID-19: perspectives of K-12 stakeholders. *EUREKA: Social and Humanities, (2)*, 61–72. <https://doi.org/10.21303/2504-5571.2023.00266>.

Baran, E. & Alzoubi, D. (2020). Human-centered design as a frame for transition to remote teaching during the COVID-19 pandemic. *Journal of Technology and Teacher Education 28*(2), 365–372.

Bethell, G. (2016). *Mathematics Education in Sub-Saharan Africa: Status, Challenges, and Opportunities*. Washington, DC: World Bank. <https://openknowledge.worldbank.org/handle/10986/25289>.

Bonsu, N. O., Bervell, B., Armah, J. K., Aheto, S. K., & Arkorful, V. (2021). Whatsapp use in teaching and learning during COVID-19 pandemic period: Investigating the initial attitudes and acceptance of students. *Library Philosophy and Practice (e-journal)* 6362. <https://digitalcommons.unl.edu/libphilprac/6362>.

Bouhnik, D., & Deshen, M. (2014). WhatsApp goes to school: Mobile instant messaging between teachers and learners. *Journal of Information Technology Education: Research 13*, 217–231. <http://www.jite.org/documents/Vol13/JITEv13ResearchP217-231Bouhnik0601.pdf>.

Cheung, A. C. K., & Slavin, R. E. (2013). The effectiveness of educational technology applications for enhancing mathematics achievement in K-12 classrooms: A meta-analysis. *Educational Research Review 9*(1), 88–113. <https://doi.org/10.1016/j.edurev.2013.01.001>.

Chirinda, B., Ndlovu, M., & Spangenberg, E. (2021). Teaching mathematics during the COVID-19 lockdown in a context of historical disadvantage. Education Sciences *11*(4), 177.

Chirinda, B., Ndlovu, M., & Spangenberg, E. (2022). Mathematics learners' perceptions of emergency remote teaching and learning during the COVID-19 lockdown in a disadvantaged context. *International Journal of Learning, Teaching and Educational Research 21*(1), 179–194.

Creswell J. W. (2013). Research design: Qualitative, quantitative, and mixed methods approaches. Sage publications: Washington DC.

Davis, F. (1989). Perceived usefulness, perceived ease of use, and user acceptance of information technology. *MIS Quarterly 13*, 319–340.

DeCoito, I., & Estaiteyeh, M. (2022). Transitioning to online teaching during the COVID-19 pandemic: An exploration of STEM teachers' views, successes, and challenges. *Journal of Science Education and Technology 31*, 340–356. <https://doi.org/10.1007/s10956-022-09958-z>.

Haleem, A., Javaid, M., Qadri, M. A., & Suman, R. (2022). Understanding the role of digital technologies in education. *A Review Sustainable Operations and Computers 3*, 275–285, <https://doi.org/10.1016/j.susoc.2022.05.004>.

Hodges, C., Moore, S., Lockee, B., Trust, T., & Bond, A. (2020). The difference between emergency remote teaching and online learning. *EDUCAUSE Review*. <https://er.educause.edu/articles/2020/3/the-difference-between-emergency-remote-teaching-and-online-learning>.

Hsieh, H. F., & Shannon, S. E. (2005). Three approaches to qualitative content analysis. *Qualitative Health Research 15*(9), 1277–1288.

Kim, J. (2020). Teaching and learning after COVID-19. Three post-pandemic predictions. Inside higher ed. <https://www.insidehighered.com/digital-learning/blogs/learning-innovation/teaching-and-learning-after-covid-19>.

Mardini, G. H., & Mah'd, O. A. (2022). Distance learning as emergency remote teaching vs. traditional learning for accounting students during the COVID-19 pandemic: Cross-country evidence. *Journal of Accounting Education 61*, 100814. <https://doi.org/10.1016/j.jaccedu.2022.100814>.

Mishra, P., & Koehler, M. J. (2006). Technological pedagogical content knowledge: a framework for teacher knowledge. *Teachers College Record 108*(6), 1017–1054.

Moonasamy, A. R., & Naidoo, G. M. (2022). Digital learning: Challenges experienced by South African University students during the COVID-19 pandemic. *The Independent Journal of Teaching and Learning 17*(2), 76–90.

Mukuka, A., Shumba, O., & Mulenga, H. M. (2021). Students' experiences with remote learning during the COVID-19 school closure: implications for mathematics education. *Heliyon 7*(7), e07523. <https://doi.org/10.1016/j.heliyon.2021.e07523>.

Mulyono, H., Suryoputro, G., & Jamil, S. R. (2021). The application of WhatsApp to support online learning during the COVID-19 pandemic in Indonesia. *Heliyon 7*((8) (Aug. 2021), Article e07853. <https://doi.org/10.1016/j.heliyon.2021.e07853>.

Nhongo, R. & Tshotsho, B. P. (2021). The shortcomings of emergency remote teaching in rural settings of Zimbabwe during COVID-19 school closures: Lessons from China's experience. *Africa's Public Service Delivery and Performance Review 9*(1), a482. <https://doi.org/10.4102/apsdpr.v9i1.482>.

Nkomo, S., & Matli, W. (2022). Emergency remote education in Southern African schools: digital transformation bridging the gap in the COVID-19 era. *International Journal of Educational Development in Africa 7*(1), 17. <https://doi.org/10.25159/2312-3540/11609>.

O'Connell, B., Tharapos, M., De Lange, P., & Beatson, N. (2022). Revitalising the enterprise university post-COVID 19: A focus on business schools. *Meditari Accountancy Research 31*(1), 141–166. <https://doi.org/10.1108/MEDAR-06-2021-1332>.

Ogbonnaya, U. I., Awoniyi, F. C. & Matabane, M. E. (2020). Move to online learning during COVID-19 lockdown: Pre-service teachers' experiences in Ghana. *International Journal of Learning, Teaching and Educational Research 19*(10), 286–303.

Raman, A., & Don, Y. (2013). Preservice teachers' acceptance of learning management software: An application of the UTAUT2 model. *International Education Studies 6*(7), 157–164.

Rogers, E.M. (1995). *Diffusion of Innovations*. 4th edition. The Free Press: New York.

Sunzuma, G., Zezekwa, N., Mutambara, T. L., Chagwiza, C., &. Gwizangwe, T. (2022). Preservice teachers' Whatsapp preferences in a mathematics methodology course during the COVID-19 *pandemic. Open Education Studies* 4(1), 225–240. <https://doi.org/10.1515/edu-2022-0014>.

Teymori, A. N., & Fardin, M. A. (2020). COVID-19 and educational challenges: A review of the benefits of online education. Annals of Military and Health Sciences Research 8(3) e105778. <https://dx.doi.org/10.5812/amh.105778>.

Thabela, T., Shumba, S., & Muntanga, D. (2020). *Education Cluster Strategy: Zimbabwe COVID-19 Preparedness and Response Strategy.* <https://reliefweb.int/sites/reliefweb.int/files/resources/zimbabwe_education_cluster_covid_strategy_12.05.2020_final.pdf>. Accessed February 2024.

Venkatesh, V. & Davis, F.D. (2000). A theoretical extension of the technology acceptance model: four longitudinal field studies. *Management Science 46*, 186–204.

Venkatesh, V., Morris, M., Davis, G., & Davis, F. (2003). User acceptance of information technology: Toward a unified view. *MIS Quarterly 27*(3), 425–478.

Venkatesh, V., Thong, J. Y., & Xu, X. (2012). Consumer acceptance and use of information technology: Extending the unified theory of acceptance and use of technology. *MIS Quarterly 36*(1), 157–178.

Weber, R.P. (1990). *Basic Content Analysis.* Newbury Park, CA: Sage Publications.

Yin, R. K. (2017). *Case Study Research and Applications: Design and Methods.* Sage Publications: Washington DC.

Yunus, M., Shin Ang, W. & Hashim, H. (2021). Factors affecting teaching English as a second language (TESL) postgraduate students' behavioural intention for online learning during the COVID-19 pandemic. *Sustainability, 13*(6), 3524.

Zhang, Y. & Wildemuth, B. (2009). Qualitative analysis of content. In *Applications of Social Research Methods to Questions in Information and Library Science* (2nd ed., pp. 318–329). Libraries Unlimited.

CHAPTER 4

# Creating Digitalized and Virtual Spaces in Real-Time Egyptian Post-COVID Mathematics Education Classrooms: Insight from the Higher Education Sector in STEM

*Mariam Makramalla*[1]
[1]NewGiza University

**ABSTRACT**
This study is situated in the Egyptian higher education sector and strategically positioned in the 'sandwich year' between the fully online and the entirely in-person teaching formats. Utilizing the inclusive pedagogy framework as a theoretical framework, the study challenges the mainstream opinion that renders the in-person instruction model more appropriate for student learning of mathematical pedagogical practice tools. This is done by shedding an advocacy lens into female postgraduate student voices that would be marginalized in the "normal setup". The data were collected from eight postgraduate students with mobility challenges. Eight longitudinal interviews were extended over one year, and an open-ended format was adopted. Data were qualitatively inductively coded. Code patterns were detected and cross-matched for accuracy via follow-up interviews. The study's results highlight the inclusive nature of online instruction to marginalized female postgraduate students. Based on these results, the chapter offers recommendations for the current postgraduate mathematics education classroom format, rendering it more inclusive to the needs of shy and highly immobile learners. Scholars are encouraged to adopt the methodological framework suggested in this study, each in their context, to achieve a wider-reaching advocacy platform for postgraduate female students in Africa.

*Keywords:* Mathematics, Egypt, female, postgraduate education

## Introduction

Despite the challenges posed by the pandemic in terms of reach and inclusion of learners, the shift to a virtual learning space has resulted in the realization of learner potential that had, until that point, been unheard of (Makramalla, 2022). This is true for marginalized learners whose physical presence in the classroom has presented a daily challenge that had been overlooked so far. Learners with a mental or physical disability, for example, had to pay a much higher price to be physically present in the same classroom with their counterparts, for whom the mere physical presence in the classroom—not only presented no challenge—but also was considered a positive re-enforcement for their learning. The same is true for learners who had to attend to caring duties. In a recent interview with a married female learner from the Middle East who was pursuing her studies in the United Kingdom, the learner indicated the following: "For me to be present in this classroom meant a huge fight with social stigma and an additionally very high financial burden when compared to my colleagues for whom coming to class was a fun learning experience. It felt very unfair to be expected to have the same focus and mental capacity when the price I am paying simply to be there is something they never had to fight with and the system never considered" (Makramalla, 2022, p. 295).

In this study, I choose to advocate for postgraduate Egyptian female learners in the field of mathematics education. The field of postgraduate study in mathematics education was particularly chosen as it substantially represents female participants (CAPMAS, 2022). Despite numbering higher than their male counterparts, the voices of female postgraduate learners have been—so far—understudied in the mainstream track investigation of Higher Education pedagogies in the African context (Alghamdi, 2021; Bezuidenhoot and Cilliers, 2010; Teferra and Altbachl, 2004). I chose to reflect on how the re-adjustments caused by the COVID virtual learning experience have opened opportunities for inclusion that have so far not been considered. I focus on the lessons from the virtual learning experience facilitated during the COVID era.

In the following, I start by situating the reader in the Egyptian context of Higher Education in a broad sense. This is followed by the presentation of statistics related to the sample chosen for this study. This quantitative account also acts as a building block to further support the rationale for choosing this participant sample. Afterwards the literature review elaborates further on previous studies conducted with the chosen participant sample. Next, I present the theoretical framework chosen for this study: the inclusive pedagogy framework. I describe my understanding of the inclusive pedagogy framework, how it is connected to the theme of this work, and how I intend to use it to serve the purposes of this investigation. The literature review and the theoretical framework are then merged to frame the research question that leads to the main investigation of this study. The second part of this chapter illustrates the overarching picture of the research design and the resulting findings. The analysis of the findings reveals interesting conclusions, which are presented in the following sections. Finally, the implications are discussed.

## Background: Postgraduate Studies in the Egyptian Higher Education Sector

To present the Higher Education sector participation rate statistically, it is important to first briefly illustrate how the K–12 education system operates to enable its graduates to be part of the Higher Education sector. The Egyptian K–12 education system is made up of two learning stages. The first learning stage is compulsory for

CREATING DIGITALIZED AND VIRTUAL SPACES

all learners (K–9), while the secondary stage (years 10–12) is optional. Around 89 % of learners who opt to pursue their secondary stage education later pursue further studies in the Egyptian Higher Education system (CAPMAS, 2022). Figure 4.1 illustrates the key learning stages in Egypt, indicating compulsory learning stages and options post-school graduation.

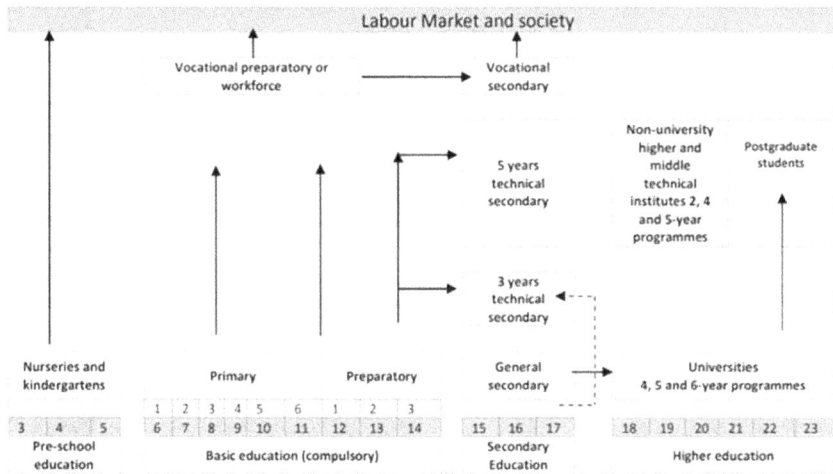

**Figure 4.1:** Egyptian K-12 sector distribution leading to higher education.
*Source:* Ministry of Education (2011), Pre-University Education System in Egypt: Background Report, Ministry of Education, Cairo.

As indicated in Figure 4.1, students who graduate from school after completing the compulsory learning stages can enter vocational education and pursue a career in the technical or service industry. The system allows students pursuing this track to pursue a vocation alongside their studies. As a result, the system allows them to become financially independent at an earlier stage than their counterparts who choose to continue pursuing their studies in secondary school in the K–12 sector.

Statistically, of the total number of enrolled students in the compulsory learning stage, less than one-third of students opt to continue their studies in secondary school (CAPMAS, 2022). Culturally, it is more acceptable for learners to pursue their studies in Higher Education (particularly undergraduate studies) right after school graduation than to pursue the job market and later re-join the Higher Education sector as undergraduates (Ead, 2019; Torres & Herrera, 2006).

Gender distribution in the higher education sector presents itself as an additional layer of complexity. Historically, patterns indicate that female students were more likely to leave school after completing their compulsory years of schooling to take up caring duties. Female students pursuing higher education usually must attend to home-caring duties alongside their studies (Zaalouk, 2013). This is particularly true of postgraduate female students in Egyptian higher education.

When considering female postgraduate learners who choose to concentrate on mathematics education, an additional layer of complexity must be considered. Mathematics as a field of study is stereotypically considered a male-dominated area (Martin, 2013). Nevertheless, statistics show that learners who pursue a postgraduate career in mathematics education in Egypt are predominantly female (CAPMAS, 2022). Hence, female learners, who statistically form the majority of mathematics education learners, are socially marginalized and considered the minority given the overarching social stigma. In addition to toggling between caring and learning duties, mathematics education postgraduate learners have also often struggled with questions of social inclusion. This triple form of isolation is one of the main reasons why this sample is worth studying and advocating for.

## The Topology of Female Students in Postgraduate Higher Education in Egypt

The overall topology of learners in Egypt's higher education sector has been presented in the previous section. In this section, I focus on female students pursuing postgraduate degrees in mathematics education in the Egyptian Higher Education Sector. The fact that this group of participants is highly understudied (Teferra and Altbachl, 2004; Zaalouk, 2013) adds to the advocacy dimension that this work aims to achieve. Statistically, less than 20% of female undergraduate students pursue postgraduate studies at Egyptian Higher Education institutions annually. Less than 5% of these students do not have other caring duties to attend to alongside their studies. In a sense, part-time or full-time female post-graduate students who toggle caring duties alongside their main study present themselves as the majority female participants in the postgraduate Higher Education sector (Hannawi and Salmi, 2018). Hence, it is remarkable that these learners' needs are rarely studied (Khozaei et al., 2015).

From a cultural point of view, sociological scholars (Faris, 2013; Makramalla, 2021; Megahed, 2013; Torres & Herrera, 2006) in the Middle East have concluded that in a cohort of male and female adult learners, female learners are less likely to openly express their opinions or debate a given topic. Female learners are also less confident presenting their work to a larger audience or engaging in a facilitated workshop session requiring them to work together in mixed-gender groupings. In a sense, female learners classified as shy or unmotivated present themselves as most of the audience in mixed-gender adult learning classrooms.

This becomes particularly relevant in the field of mathematics education as the nature of the pedagogical training requires students to be comfortable with an interactive discussion and activity setting. I choose to focus on the intersection between these two understudied constructs.

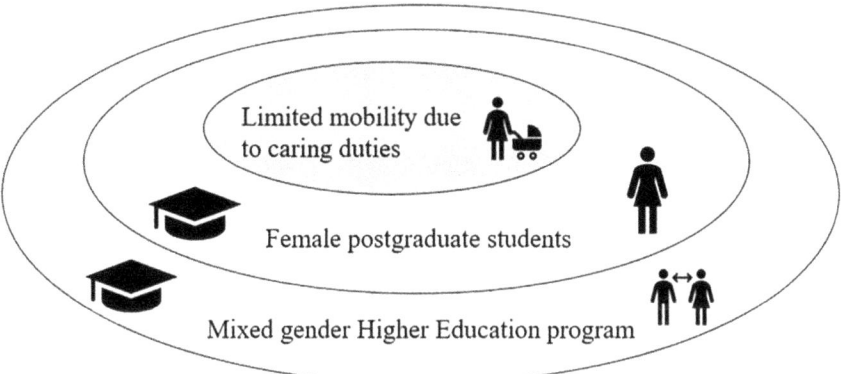

**Figure 4.2:** Sample for the study.

As illustrated in Figure 4.2, within the wider framework of a mixed gender Higher Education program, I choose to particularly focus on postgraduate female students pursuing their degrees in education. Statistically, most of these students have caring duties to attend to alongside their studies. My sample of choice is, hence, postgraduate female students at the faculties of education (Department of Mathematics Education) have limited mobility due to home caring duties. It is important to mention that these caring duties extend beyond looking after children. Despite being the most prevalent source for caring duties, some of the participants interviewed were looking after aging family members or partners with physical or mental disabilities. The Research Design section discusses more details about the sample of choice.

## Theoretical Framework: The Inclusive Pedagogy

The inclusive pedagogy framework builds on the foundation that an educational system needs to be channeled to target the whole development of the learner. Accordingly, the different layers of complexity in learner status, well-being, history, and culture need to be incorporated into the creation of a given pedagogical experience (Spratt and Florian, 2014). In a sense, it could be claimed that for an educational experience to target the whole person, it would ideally need to be tailored for each student in a different way. This claim, however, on the outside, seems to challenge the mainstream claim of achieving equality. Equality in a pedagogical experience would indicate that all students have equal rights to access a given learning experience. Scholars (Florian and Black-Hawkins, 2011; Salmi and D'Addio, 2021) have long debated this apparent paradox between the equality of learning experiences and the uniqueness of student complexity. This paradoxical realization leaves us with the two questions that on the outside seem to be contradicting and that Minow (1990, p. 72) reflects on in his work:

1. When does treating people differently emphasize their differences and stigmatize or hinder them on that basis?
2. When would treating people the same become insensitive to their differences and hence more enforcing existing stigmas?

In response to this paradox, scholars (Spratt and Florian, 2014) have suggested the inclusive pedagogy framework. The ethos of the inclusive pedagogy framework counterbalances preexisting frameworks by extending additional inclusive techniques to curricula that are equally available to all students. In other words, instead of making a different curriculum available to students with some contextual needs, inclusive pedagogy targets the extension of what is originally available to all students. The inclusive pedagogy framework, hence, rejects preexisting claims that students have a fixed ability to learn things in a certain way (Haumovitz and Dweck, 2017). Instead, it utilizes current learning experiences as an experimentation space to better identify and predict the most productive channels by which to reach out to each student. The inclusive pedagogy framework is hence based on the foundation that a student's capacity is changeable. The approach adopted by an instructor would, therefore, shape the way in which a student thinks about a concept or method of choice. In a sense, the inclusive pedagogy framework rejects the notion that learning is neutral. It extends its scope to adjust the learning channel based on predictions gathered and studied in real-time from the student cohort. Figure 4.3 illustrates my perception of the theoretical framework.

In Figure 4.3, I aimed to map my understanding of the inclusive pedagogy framework. Building on the common understanding of fair and equal education, which aims to ensure equality of access and provision, the inclusive pedagogy framework presents itself as a potential solution to the dilemma of difference posed in scholarship (Norwich, 2009). By extending the existing equally available curriculum, the aim is to be inclusive of students with special needs without stigmatizing them as different.

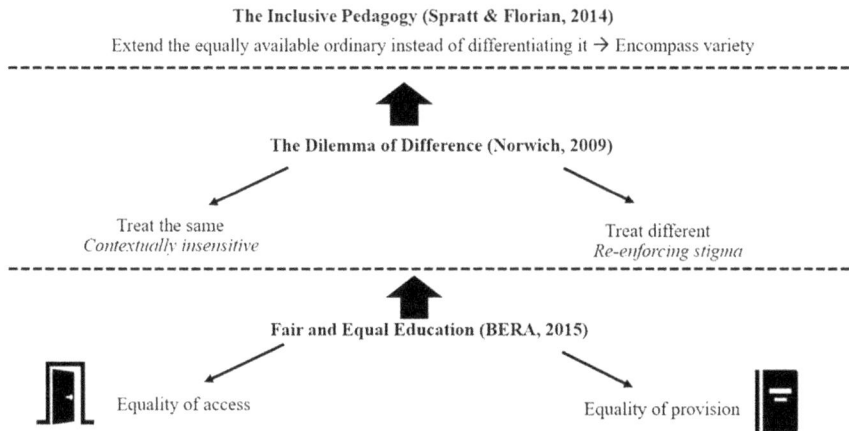

**Figure 4.3:** Inclusive pedagogy framework: conceptualization for this study.

The target of this study was to explore how existing pedagogical frameworks (Levy, 2007; Ozdamli, 2012) often seem to overlook the needs of some of their stigmatizing participants (such as female students attending to caring duties). Existing theoretical frameworks that have aimed to be inclusive of all students often suggest "add-on" components to the existing pedagogical framework so that the needs of non-mainstream students can be met (Levy, 2007). In a sense, it could be argued that such an approach would be pinpointing the differences between students instead of blending these into the mainstream. Instead of highlighting their special needs as different from the mainstream (Busher et al., 2013), this study aims to propose an inclusive approach that would challenge the existing pedagogical frameworks. By suggesting an expansion from within, the inclusive pedagogy framework presents itself as a theoretical framework that can cater to a larger base of student audience in a way that would render each participant with the sense of being part of the system and not an exception to the system. For the scope of this study, the students with special needs are the female postgraduate students at the faculties of education (with a concentration on mathematics education) who have access and mobility problems resulting from their caring duties. With this theoretical framework as an underpinning block, this chapter aims to address the following research question:

> *How can we capture the lessons learnt from the virtual learning topology of postgraduate instruction during COVID in order to suggest an extension to the regular postgraduate instruction model in the Egyptian Higher Education sector that would be less stigmatising of female students with mobility problems?*

The next section presents the research design, which includes the methodology and analytical framework that have been adopted in response to this question.

## Research Design

This study adopts an embedded single case study design methodology (Yin, 2009). Figure 4.4 presents my overarching illustration of how this methodology was adopted for the current study.

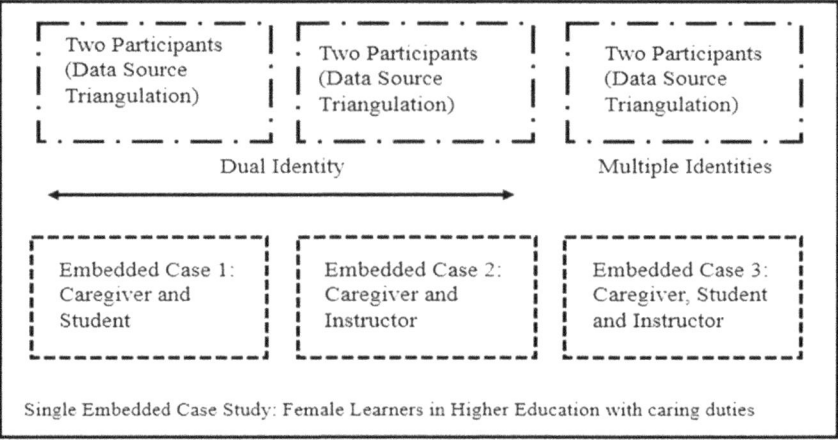

Figure 4.4: Single-embedded case study design.

As illustrated in Figure 4.4, the study adopts the case of marginalized female students as the wider context of the case study. Embedded in this wider case study are a number of marginalized typologies, all of which relate to postgraduate female students. Each of these typologies was represented by two participants for data source triangulation purposes (Yin, 2009). The data collection took the form of extended semi-structured interviews, whereby participants were encouraged to reflect on their dual or multiple identities (student–instructor; caregiver–student; instructor–caregiver).

In this section, the aim is to present in more detail how the aforementioned choice of participants has been investigated to serve the purpose of responding to the research question at hand. I start by reiterating the rationale for the study, highlighting how this has informed both sample selection and data collection methodology. This is followed by an overarching presentation of the data collection protocol, outlining the data collection methods and their validity and reliability.

Finally, I conclude by presenting the analytical framework that has been adopted. This leads up to the findings, which are separately presented in the section to follow.

## Rationale for the Study

As already discussed, the rationale for the study is to explore the degree to which female students with caring duties related to mobility problems are disadvantaged by the existing mainstream instructional model. This study derives its relevance from the fact that this category of participants forms the mainstream baseline of the audience for postgraduate faculties of education (Hussein, 2016). Despite these life circumstances presenting themselves as a problem that most students must deal with, it is remarkable to see how this matter remains understudied (Ong et al., 2017). In addition to practically informing the relevant stakeholders, this study also presents itself as an advocacy platform for this marginalized group.

## Sample

The sample chosen for this study included eight postgraduate female mathematics education students with mobility problems due to maternal caring duties and caring duties related to other more senior family members to whom the participant of the study was the sole caregiver. Most participants in the study also were toggling a full-time teaching job alongside their full-time and/or part-time postgraduate studies at the faculty of education (Figure 4.4). This means that, as participants were reflecting on their experiences as students, they were also reflecting on how this informed their role as instructors in dealing with other students who are often faced with other mobility problems. In a sense, students who were struggling with their own inclusion in the higher education pedagogical framework were also members of this framework in their roles as instructors. As instructors, participants of this study had their own student cohorts that often were themselves also females attending to caring duties (Figure 4.5).

As evident in Figure 4.5, the intention for the selection of this particular sample was that two-thirds of the sample participants would be able to reflect on their marginalization from the perspective of their dual reality. In a sense, two-thirds of the sample struggled with their own sense of marginalization while enforcing it on similarly marginalized students as part of their dual role in the system. The third group of participants was a group that carried a dual identity that would enable them to reflect on their sense of marginalization. Yet, they were not part of the system and offered a fresh perspective on how they would want to expand the framework. The reflections of all participants were analyzed

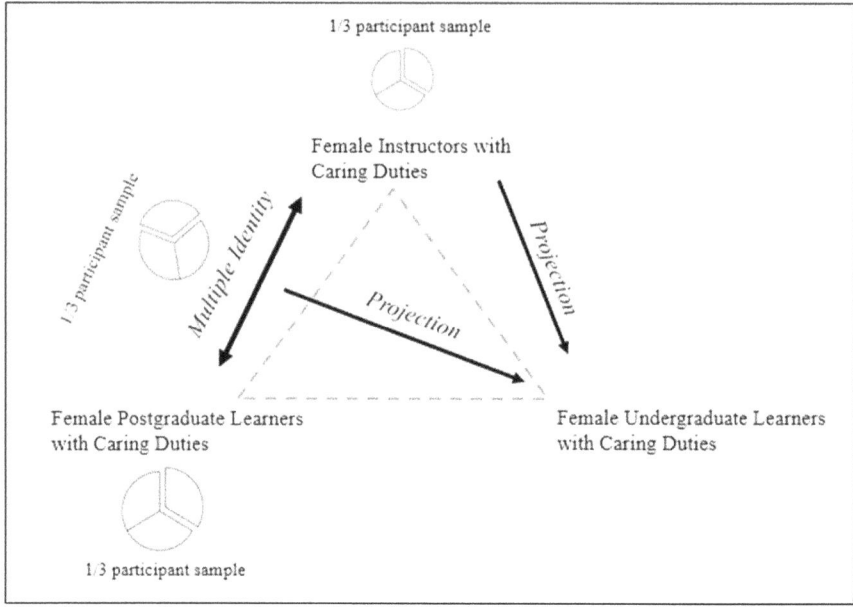

**Figure 4.5:** Sampling layout.

from the perspective of the inclusive pedagogy framework, which will be further outlined in the next sections. Table 4.1 presents the abbreviations that will be used when referring to each of the members of the sample.

**Table 4.1:** Abbreviations of the members of the sample

| Sample | Number of Participants | Abbreviation |
|---|---|---|
| Caregiver and Student | 2 | CL1, CL2 |
| Caregiver and Instructor | 2 | CI1, CI2 |
| Caregiver, Student and Instructor | 2 | CLI1, CLI2 |

The following section outlines the data collection protocol adopted with the aforementioned sample groups of participants.

## Data Collection Protocol

The data collection protocol consisted of an extended open-ended interview that was replicated eight times with each of the participants. The interview was mainly made up of five sections, which are presented in Table 4.2.

Table 4.2: Open-ended interview structure

| Section | Content | Target |
|---|---|---|
| 1 | Overview of the context of the study Discussion about confidentiality and other ethical restrictions | The target of this section was to familiarise participants with the wider work that this interview was contributing to. Participants were briefed about the wider research and were offered the chance to ask questions. Participants were also consented for using their interview contributions for research purposes. |
| 2 | Introduction of the participant | Participants were given the opportunity to introduce themselves in their dual roles as instructors and/or postgraduate students. Participants were offered a set of questions that formed a self-reflection space for participants to relate to their multiple identities (as women, caregivers, students, teachers etc.) |
| 3 | The COVID Instruction Mode from the perspective of being a postgraduate student | During a set of open-ended questions, participants were encouraged to reflect on their experiences in the distance learning mode that was necessitated due to the pandemic. Participants were encouraged to think of their whole-person identity that was established in section 2 as they presented their responses to this section. |
| 4 | The COVID Instruction Mode from the perspective of being an instructor | Building on the whole-person self-identification approach that has been established in section 2, participants were encouraged to view their students (if available) from this same perspective. Building on their own experience as distance students, participants were now encouraged to discuss how this has informed their roles as distance learning instructors. |
| 5 | Clarification Questions | The interview was concluded with clarification questions that related to matters that arose during interview sections 2-4. |

It is out of the scope of this chapter to go into a detailed account of the open-ended questions and how they served the purpose of each section. It is important to mention that the questions have been designed with the theoretical framework as a guide. The inclusive pedagogical framework offers a solution to the dilemma of difference that scholars have discussed (Minow, 1990; Norwich, 2009). Extending the existing framework offers a differentiated mode of instruction that is contextually sensitive while not re-enforcing existing social stigma. In my attempts to explore the potential extension of the Higher Education pedagogical framework, the target of the extended interviews was to explore how participants from the system (often with dual or multiple roles) related pedagogically to the digital mode of instruction.

It is also out of scope to present a detailed account of the validity and reliability of the data collection process. For the scope of this chapter, it is worth mentioning that Section 5 (Table 4.2) has been added to the data collection protocol to establish an element of redundancy, thereby strengthening the reliability of the study (Yin, 2009). Multiple sources were interviewed for each of the embedded case studies to strengthen the validity of the study (Yin, 2009). The next section outlines how the interview data were coded and analyzed.

## Analytical Framework

The analysis of the interview script was underpinned by inclusive pedagogy as a guiding framework. Participants' responses that seemed to be offering input about a potential expansion of the existing system and hence were considered to be aligned with the underpinning theoretical framework were coded according to an inductive data coding format (Yin, 2009). Each section of the interview revealed data points that were inductively coded. As the inductive process evolved, patterns emerged between the different data points. This was either because a certain theme (denoted by a code) was repeatedly referred to, often in different ways, or because certain connections were made between data points that had originally seemed independent of each other. Hence, as illustrated in Figure 4.6, patterns emerging from the data were each separately coded. Codes were then mapped together in terms of (1) their frequency of appearance and (2) their interrelatedness. Figure 4.6 visually depicts the approach I have adopted regarding data coding.

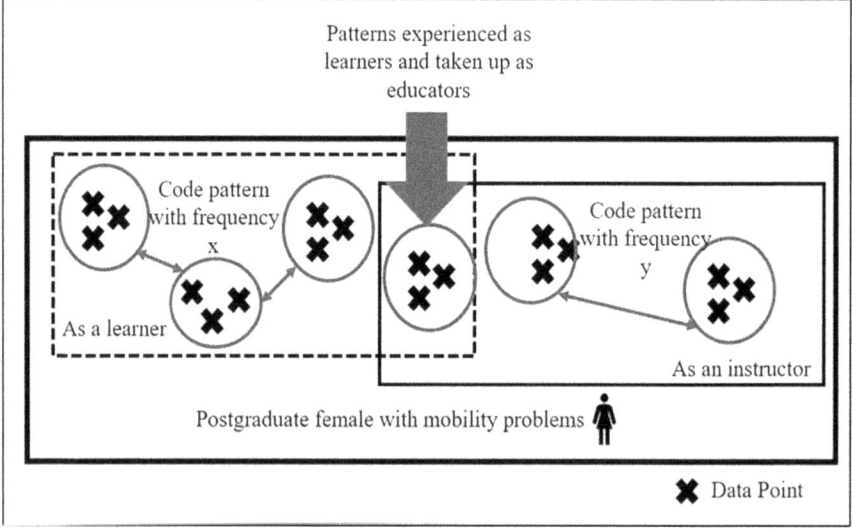

**Figure 4.6:** Analytical approach: inductive data coding.

As illustrated in Figure 4.6, repeating patterns were considered of high importance, and follow-up questions (see Table 4.2; Section 5) were designed to investigate these further. Emerging data points (codes) were also mapped from a relational point of view. For this particular investigation, the mapping was mostly causational, showcasing why participants related to a given matter in a given way how different data points can be connected to each other and the identity of the student in question. This latter form of mapping (inter-relativity) has mostly informed the attempt to respond to the research question at hand.

**Table 4.3:** Example of inductive data coding

| Data Extract | Participant | Coding |
| --- | --- | --- |
| "When I was doing my own assignments I noticed how helpful it was to use digital resources for statistical mapping of my findings. I had never used this before. I started using this for visually plotting out the learning stages of my own students" | CLI 1 | Use of statistical tools for instructional purposes |

Table 4.3 presents an illustrative extract from the interview logs and shows how this was qualitatively inductively coded as part of the data analysis. As the data were inductively coded, similar extracts were mapped together under the same data coding point. A distinction was made between the sources of a given data coding point. If participants from different sample groups seemed to agree on a given point, this data coding point was considered a multi-node. If the input was coming from only one sample group, then this coding point was considered a single node. Nodes were then connected to each other to form data clusters.

While this analysis can serve multiple purposes, I choose to scope my presentation of findings to the codes that intersect with the dual identity of the participant. In other words, for the scope of this work, I choose to present learning practices that the student appreciated to the extent that these have become part of their own practices as an educator. In presenting these findings, I build on the inclusive pedagogy framework in showcasing how lessons learned from the distance learning format necessitated by COVID and how this can lead to the extension of daily practices to cater to currently underserved needs of female students (in alignment with the inclusive pedagogy framework). The next section presents an account of some code clusters that I chose to focus on for the scope of this chapter.

## Findings: Patterns Identified as Potential Expansion Points to the Existing Framework

As stated earlier in this chapter, the postgraduate study of classroom topology in mathematics education presents a social paradox. On the one hand, it is densely populated with female students (Khozaei et al., 2015). On the other hand, it is stereotypically affiliated with male students (Martin, 2013). The target of this work was to shed light on the stereotypical affiliations and to focus on advocating for the voices of the existing mainstream population of students, namely those of female students who often carry a double role as caregivers and students and/or as students and instructors.

The findings of this study revealed three very interesting multi-node patterns that were unanimously agreed upon by all three participant groups (CL, CI, and CLI). These patterns denote pedagogical trends that have indirectly resulted from the distance learning mode of postgraduate instruction. These trends have been repeatedly referred to as being very conducive to learning. They are hence being suggested as an expansion of the existing pedagogical framework in alignment with the inclusive pedagogy framework. In the following, these three trends are presented as reported by all three participant groups (CL, CI, and CLI). The next sections also present how two participant groups (CI and CLI) attempted to adopt these trends in their dual identity as instructors of female students who were also attending to caregiving duties.

## Findings Pattern 1: Promoting Access and Availability of Resources

The distance learning format has facilitated the availability of learning resources beyond the synchronous space of the actual classroom time. In other words, in most cases, instructional sessions were recorded, thus allowing participants to re-listen to the session recording at a time that was suitable for them. This function was previously unfamiliar to most postgraduate students (CL, CI, and CLI) in the target group, who had to be synchronously and physically present to receive a given lesson.

This sense of flexibility regarding both time and access has provided the target group with two benefits. Firstly, they were able to schedule their learning in accordance with their other duties. This flexibility of scheduling recognizes the agency of postgraduate students, which in turn motivates them to take ownership of their learning. One participant (CL2) phrases this in the following way:

> *I feel that I am treated as an adult. The system acknowledges my responsibilities, and I feel I am trusted to design my learning experience for myself in a way that aligns with my whole person.*

Secondly, the recorded format meant that students had the possibility to re-listen to parts of the sessions that they might have missed as they were attending to their caring duties. Students who were synchronously attending a given session but had to toggle while other caring duties were provided with a buffer space, as they realized they could always catch the parts they were missing. This meant that students could always focus on the present, since they were assured that they could catch up on what they might have missed in parts of the session that had happened in the past. In this regard, one participant (CLI1) expressed the following:

> *From a mental health point of view, it is very relieving to see that I don't have to stress about the parts I missed because I can always listen to them later. This way, I can focus on the parts being discussed now.*

The availability and access of learning materials in an asynchronous format have led to the establishment of an instructional habit, which is that interview participants (CI1, CI2, CLI1, and CLI2) have repeatedly been referred to as "the online library." I present this expanded pedagogical tool in the next sub-section.

*The Online Learning Library.* Due to their growing appreciation for learning resources, participants (CI1, CI2, CLI 1, and CLI 2) have resorted to creating their own—so-called online libraries—where they keep all the teaching resources (and often recordings) that they have utilized during their synchronous in-person instruction for students to use in their own study time.

> *I have started this trend by making all the resources I use available for them [the students]. I even uploaded some recorded videos where I explained the main concepts. This can help them [the students] be re-assured that they can tap into support materials in case they missed parts of my explanation in class.* (CLI2)

The online library compiles resources for students to refer to during study time. Considering the inclusive pedagogy framework, this extension of the learning experience means that instruction can be more inclusive for slow students or students with mobility or mental health problems.

## Findings Pattern 2: Increased Utilization of Digital Statistical Tools

Prior to the distance learning mode that the pandemic has necessitated, statistical tools for assessing student knowledge acquisition had not been automatically connected to the regular assessment tools for testing student mastery of a given skillset or competency. Students used to be primarily assessed in a summative paper-based format. At the same time, paper surveys would serve the purpose of assessing their learning experience. Participants of the interview (CI1 and CI2) repeatedly expressed that the online mode of assessment created a natural link between statistical analysis and student assessment that had previously not been considered. One participant (CI2) expressed this in the following way: *"I can now test my students while keeping an eye on their overall performance and improvement, both at the same time."*

Through the digitalization of the assessment process, participants are now equipped with the capacity to statistically track and monitor their students' development. This is a by-product of the digitalized teaching and learning experience that participants (CI1 and CI2) suggest as a potential expansion to the existing pedagogical framework.

Consequently, three participants (CI1, CI2, and CLI1) have resorted to broadening their perspective of the value assessments they served, as will be further elaborated in the next section.

*Extension of Assessment Typology.* In alignment with ongoing research (Zaalouk et al., 2021; Makramalla, 2022), Egyptian teachers are even more realizing the importance of transferring their experience of toggling between different roles that they can acquire, while attending to their day-to-day roles in their mathematics classrooms. In the classroom, they are not only teachers. Their roles are extended to becoming assessors of the overall students' learning experience. This extension of the teachers' roles ensures sustained quality and improvement.

As highlighted by the participants of this study, the utilization of online assessment tools, which automatically translate into statistical units, enables the participants to expand their regular assessment framework. From the perspective of the inclusive pedagogy framework, the extension of the assessment framework to include digital tools provides participants with a dataset that can serve multiple purposes. Assessment data can serve a purpose that goes beyond the testing of student ability. Assessment data can serve the participant to expand their role beyond that of an instructor to acting as a stationed in-class action researcher who seeks to perfect the teaching and learning experience.

## Findings Pattern 3: Empathy-based Approach

The multi-node code that most frequently emerged during the inductive data coding process was the one that related to a sensation of a more empathy-embedded approach to learning. Participants (CL1 and CL2) repeatedly commented that the pandemic caused more widespread problems and made their instructors more aware of the multifaceted reality of each student. One participant (CL2) stated this in the following way:

> *In the past [prior to COVID], we had to hide our problems as this seemed unprofessional [to display their full person in class]. In class, I had to display only part of my personality that had to do with being a student. This was different during COVID.*

This empathy-based approach to teaching and learning seemed more natural and inclusive to participants (CI1 and CI2) who acted as postgraduate students and decided to expand their existing pedagogical framework to include this dimension in their educator roles. The next section elaborates on this.

*Redefined Perspective on Vulnerability.* As an extension to the already existing pedagogical framework, participants have realized the importance of seeing the student as a whole person. This whole person is naturally affected by challenges that happen outside of the classroom. Hence, there is a direct connection between a student's performance and the wider challenges that s/he must deal with as a whole person. As a result of this realization, participants have re-defined their perspectives when it comes to providing their students with the space to be open about their vulnerabilities. This space for openness about one's vulnerability is facilitated through the openness of teachers to listen and engage with their students beyond their academic needs.

The adopted data collection approach that views the marginalized students as the main data collection point informing an expansion of pedagogy was reported to be very empowering to the participants, especially those with a dual role in the Higher Education sector (CI1, CI2, CLI1, and CLI2). It also renders an expanded pedagogical framework more accurate in its inclusivity, as its pillars are derived from the non-mainstream participants themselves. The next section discusses how this process can be adopted to achieve more advocacy-based expansions to pedagogical frameworks in other African contexts. I also describe how, in particular, the expansion into the digital learning spaces can be considered relevant to a larger audience from an inclusive pedagogical framework stance.

## Discussion and Implications

This work has been underpinned by the inclusive pedagogy framework, which suggests expanding the existing teaching and learning pedagogy to be more inclusive to an otherwise marginalized target group without directly stigmatizing them as different. The chosen target group for this study is female postgraduate mathematics education scholars with limited physical mobility. The study is timely, situated directly after the distance learning era necessitated by the pandemic. The research question mainly explores lessons that the participant group has learned during COVID and that have informed their practice as educators.

In summary, three main lessons have been presented. Firstly, questions of availability and access to resources beyond the synchronous and physical classroom space were discussed. Participants expressed how this access has empowered their sense of agency and, as a result, started the trend of making a digital library available for their students. Secondly, questions of connectedness between student assessment and statistical analysis were highlighted. Participants indicated how the digital mode of assessment had equipped them with the capacity to extend their roles in the classroom to go beyond a mere teaching role. As a result of this realization, participants extended their assessment topology to include a digital dimension to facilitate their continuous improvement potential. Finally, questions of empathy have become less unmentionable. Participants have realized the importance of embedding empathy in their philosophy as educators.

This study acts as an advocacy platform, shedding light on the whole-person reality of every student and advocating for the need for this reality to be acknowledged. The caring duties of female postgraduate students cannot be set aside as separate from their identity as scholars, instructors, and students. This study revealed that as this reality gets practiced by the institutions that govern the group's instruction, they themselves—as educators—mirror the appreciation of this validation by embedding similar foundations as part of their own teaching and learning practice. This cascading of a re-envisioned philosophy (Torres & Herrera, 2006) results in a more inclusive pedagogy, where extensions to the existing curricular framework have been implemented to ensure that students with unmet needs are acknowledged and well-catered for.

Even though these findings have been connected to a very specific context (the Middle East) and a very specific participant group (females with mobility problems), both the methodology adopted for investigation and the lessons learned are extendable to other contexts of scholarly work. The methodological approach adopted to explore the opinions of this marginalized group can easily be extended in a study with other groups that are marginalized in different ways and for different reasons (Yin, 2009). Additionally, the lessons learned that have been presented as part of the findings of this study can easily be expanded to other contexts beyond the Middle East.

Beyond expanding the existing pedagogical framework in the higher education sector to adopt three digital perspectives that would facilitate access to marginalized groups, this study calls on other African scholars to adopt a similar grass-roots-based methodology built on a theoretical framework that does not re-enforce the stigma to investigate the needs of other marginalized stakeholders in the higher education sector.

The implications of this study are relevant for curriculum writers, educational policymakers, and non-profit organizations that aim to empower women. By building the analytical investigation on a ground-up inductive approach, the target was to base the recommendations on the experiences of members from this marginalized group, thus rendering them more reliable and extendable.

## Bibliography

Alghamdi, A. (2021). COVID mandated self-directed distance learning: Experiences of Saudi female postgraduate students. *Journal of University Teaching and Learning Practice 18*(3), 124–136.

Bezuidenhout, A., & Cilliers, F. (2010). Burnout, work engagement and sense of coherence in female academics in higher-education institutions in South Africa. *Journal of Industrial Psychology 36*(1), 121–134.

Busher, H., James, N., Piela, A., & Palmer, A. (2013). Transforming marginalised adult learners' views of themselves: Access to higher education courses in England. *British Journal of Sociology in Education 35*(3), 800–817.

CAPMAS. (2022). *The Statistics Relating to Higher Education and Female Education* (Open Access Publication). Central Agency for Public Mobilisation and Statistics.

Ead, H. (2019). Globalisation in higher education in Egypt in a historical context. *Research in Globalisation 1*(1), 87–93.

Faris, D. (2013). *Dissent and Revolution in a Digital Age: Social Media, Blogging and Activism in Egypt.* London: I.B. Tauris.

Florian, L., & Black-Hawkins, K. (2011). Exploring inclusive pedagogy. *British Education Research Journal 37*(5), 813–828.

Hannawi, S., & Salmi, I. (2017). *Time to Address Gender Inequalities Against Female Physicians.* Oman: Wiley.

Haumovitz, K., & Dweck, C. (2017). The origins of children's growth and fixed mindsets: New research and a new proposal. *Journal of the Society for Research into Child Development 88*(6), 1849–1856.

Hussein, H. (2016). The effect of blackboard collaborative based instruction on pre-service teachers' achievement in the EFL teaching methods course at faculties of education for girls. *Journal of English Language Teaching 9*(3), 49–67.

Khozaei, F., Naidu, S., Khozaei, Z., & Salleh, N. (2015). An exploratory study of factors that affect the research progress of international PhD students from the Middle East. *Journal for Education and Training 57*(4), 448–460.

Levy, M. (2007). Culture, culture learning and new technologies: Towards a pedagogical framework. *Journal of Language, Learning and Technology 11*(2), 104–127.

Makramalla, M. (2021). Why do we learn? A public engagement project for Egyptian schools. In Marks, R. (Ed.). *Proceedings of the British Society for Research into Learning Mathematics 41*(2), 356–362.

Makramalla, M. (2022). Redefining distance learning for the African context: Lessons learnt from Egyptian educators. In Chirinda et al. (Eds.), *Mathematics Education in Africa: The Fourth Industrial Revolution* (pp. 293–307).

Martin, D. (2013). Race, racial projects and mathematics education. *Journal for Research in Mathematics Education 44*(1), 316–333.

Megahed, N. (2013). Towards math-based architectural education in Egyptian Engineering Faculties. *Nexus Network Journal 15*(1), 565–581.

Minow, M. (1990). *Making All the Difference: Inclusion, Exclusion and American Law*. London: Cornell University Press.

Norwich, B. (2009). Dilemmas of difference and the identification of special educational needs and disabilities: Internal perspectives. *British Educational Research Journal 35*(3), 447–467.

Ong, A., Smith, J., & Ko, L. (2017). Counterspaces for women of color in STEM higher education: Marginal and central spaces for persistence and success. *Journal of Research in Science Technology 55*(2), 206–245.

Ozdamli, F. (2012). Pedagogical framework of m-learning. *Journal for Social and Behavioral Sciences 31*, 927–931.

Salmi, J., & D'Addio, A. (2021). Policies for achieving inclusion in higher education. *Policy Reviews in Higher Education 5*(1), 47–72.

Spratt, J., & Florian, L. (2014). Developing and using a framework for gauging the use of inclusive pedagogy by new and experienced teachers. *Journal for Measuring Inclusive Education 3*(1), 1479–1484.

Teferra, D., & Altbachl, P. (2004). African higher education: Challenges for the 21st century. *Journal for Higher Education 47*(1), 21–50.

Torres, C., & Herrera, L. (2006). *Cultures of Arab schooling: Critical Ethnographies from Egypt*. State University of New York Press.

Yin, R. (2009). *Case Study Research: Design and Methods*. London: SAGE

Zaalouk, M. (2013). *Globalisation and Educational Reform: What Choices for Teachers*. Routledge.

Zaalouk, M., El-Deghaidi, H., Eid, L., & Ramadan, L. (2021). Value creation through peer communities of learners in an Egyptian context during the COVID-19 pandemic. *International Review of Education 67*, 103–125.

CHAPTER 5

# Strengths, Challenges, and Implications of Digital Pedagogy for Mathematics Education: Exploring South African Postgraduate Students' Experiences

*Jayaluxmi Naidoo[1] and Rajendran Govender[2]*
[1]University of KwaZulu-Natal
[2]University of Western Cape

**ABSTRACT**

Higher Education Institutions swiftly embraced digital pedagogy in the twenty-first-century era to be future-ready. Moreover, during the 2019 Coronavirus disease (COVID-19) pandemic, teacher education institutions robustly navigated towards digital pedagogy. This chapter reports on a study focusing on postgraduate students' experiences of digital pedagogy for mathematics education. This qualitative study was located at a Higher Education Institution in South Africa and was framed by the Substitution, Augmentation, Modification, and Redefinition Theory Framework (SAMR) model. The SAMR model is an influential conceptual tool for integrating digital educational tools into pedagogy. Postgraduate mathematics education students were invited to join this study. Data generation included interactive workshops and discussion forums conducted via different digital platforms, such as WhatsApp, Moodle/Learn 2021, Microsoft Teams, and Zoom. The findings exhibit the strengths, challenges, and implications of digital pedagogy for mathematics education. This study's results reveal a need for preparation workshops to use digital pedagogy successfully in mathematics. Also, although learning time is extended when using digital pedagogy, access to essential digital resources, digital tools, and active collaborative engagement within digital platforms is necessary for effective teaching and learning. This study adds to the evolving knowledge concerning digital pedagogy for mathematics education in developing countries.

*Keywords:* Digital pedagogy, Hovercam, mathematics education, SAMR

## Introduction

The 2019 Coronavirus disease (COVID-19) pandemic encouraged Higher Education Institutions globally to advance rapidly in digital teaching, learning, and assessment. Digital pedagogy, a strategy for teaching and learning using digital platforms and digital tools, was considered adequate for preventing the spread of the contagious COVID-19 virus (Murgatrotd, 2020). Digital pedagogy in this study is the use of digital pedagogic tools and digital pedagogic platforms for teaching, learning, and assessing. These digital pedagogic tools include the Hovercam,[1] mobile phones, and computers. Digital pedagogic platforms include

---

[1] The Hovercam is a portable device connected to the lecturer's computer. The lecturer can use the Hovercam to display and project information and material in real time as the lecture progresses. The Hovercam can also be used to enlarge pictures, resources and images as the lecture progresses. The Hovercam may be used to record lectures.

Zoom, Microsoft Teams, Moodle/Learn 2021[2] and WhatsApp. Within the ambits of this study, digital pedagogy provided the students and lecturers with various innovative methods of communicating and transmitting knowledge.

Moreover, in the Fourth Industrial Revolution (4IR) era, it is important to adapt to digital pedagogy and use innovative online teaching, learning, and assessment tools (Sailin and Mahmor, 2018). Digital tools, devices, and platforms are evolving quickly and can support traditional pedagogy in many educational environments (Naidoo & Govender, 2021). The 4IR educational environments encourage pedagogy, which promotes critical thinking, collaboration, interactive technology-based teaching, learning, and problem-solving methodologies while integrating digital tools and digital platforms (Boholano, 2017; Buzzard et al., 2011; Goertz, 2015). In addition, in this study, it was apparent that digital pedagogy promoted the effective teaching and learning of skills and content. Thus, online teaching, learning, and assessments were effectively guided. Also, for this study, digital platforms and web-based resources were used with mobile phones or computers, which supported teaching and learning using videos, audio files, and text (Peachey, 2017). Digital pedagogy assists lecturers in using digital resources to upload teaching and learning materials onto digital platforms and promotes sustainable learning (Nanjundaswamy et al., 2021). This study aimed to respond to the key research question: What are postgraduate students' experiences of digital pedagogy for mathematics education?

## Twenty-First Century Teaching, Learning and Mathematics Education

Universally, teaching and learning within the twenty-first century have required a transformation in educational environments (Boholano, 2017). In the 4IR, the Internet of Things (IoT), technology, virtual reality, robotics, and artificial intelligence (AI) are widespread (Pyper, 2017). Therefore, it is apparent that technology plays an important role in transforming education worldwide (Gilbert, 2015). For mathematics education, a student's ability to communicate efficiently, collaborate, problem-solve, and think creatively and critically is important for advancing in the 4IR era. In addition, students require more time to reflect on and work with content material, enhancing student achievement (Kidron and Lindsay, 2014). Moreover, the mathematics education curriculum must link

---

[2] The participating university changed the name of its official online learning platform from Moodle to Learn 2021 during the COVID-19 pandemic.

mathematics content knowledge to real-world knowledge. Mathematics teaching, learning, and assessment should be relevant and realistic for students so that the link between what they learn and how it relates to the world around them is apparent (Van den Heuvel-Panhuizen and Drijvers, 2020).

The need for Information Communications and Technology skills has increased in mathematics education, and lecturers in mathematics education ought to adapt to this need (Bescherer, 2020). Students must be familiar with learning using digital pedagogy, tools, and platforms. Not being familiar with or not having adequate training creates obstacles to teaching and learning using technology (Palaigeorgiou and Grammatikopoulou, 2016). Thus, lecturer proficiency in digital pedagogy, tools, and platforms must be developed (Nanjundaswamy et al., 2021). Mathematics activities integrating digital pedagogy and tools encourage students to progress as they interact, participate, and collaborate on a globally linked digital platform (Alabdulaziz, 2021). Moreover, the use of Moodle for teaching, learning, and assessments of mathematics in Higher Education scaffolds students' learning by stimulating students' interest and enhancing students' performance (Handayanto et al., 2018; Naidoo, 2020).

## Digital Pedagogy, Tools and Platforms

In the era of the 4IR, using digital pedagogy, tools, and platforms is widespread (Gilbert, 2015; Nanjundaswamy et al., 2021). Within the 4IR, globally, educational institutions are inclining towards using digital tools and devices for teaching and learning (Ghavifekr et al., 2016; Moloi and Mhlanga, 2021; Sailin and Mahmor, 2018). Digital pedagogy integrates digital tools, devices, and resources in educational contexts to support online teaching, learning, and assessment. As discussed in this chapter, digital devices, tools, and resources also include computers or other devices that work with text, audio, video, and pictures to enhance learning, teaching, and assessment (Peachey, 2017).

Using digital pedagogy has increased as digital tools and devices become more affordable (Rasanen et al., 2019). To use digital pedagogy effectively, lecturers must choose the appropriate digital devices to match the objective of their pedagogy. This was evident in the study under focus, as discussed in the Findings and Discussion section. Lecturers in this study used online platforms and tools that were accessible to their students. They used Zoom, Moodle, Microsoft Teams, mobile phones, and WhatsApp to support their pedagogy.

Consequently, their pedagogy was focused on their students' needs. Thus, digital devices should be considered an important part of teaching, learning, and assessment. Various digital resources, applications, websites, software platforms, and programs may enhance digital pedagogy (Ghavifekr et al., 2016; Handayanto et al., 2018; Naidoo & Govender, 2020; Nanjundaswamy et al., 2021; Pope and Mayorga, 2019). Due to the variety available, it is not easy for lecturers to select the most applicable digital tool or resource. The notion is to choose a digital device or resource that is affordable and most accessible to students, and that will also support and enhance teaching, learning, and assessment. The lecturer's task is to reflect on teaching and learning (Chang, 2019) and encourage the use of digital tools and devices to advance education and achieve success in the lecture room (Montrieux et al., 2015). Thus, designing lectures that incorporate digital tools and devices must be carefully considered to ensure that students' understanding is promoted (Drijvers, 2013; Nanjundaswamy et al., 2021). This was evident in this study, as discussed in the Findings and Discussion section. For example, the recordings of the workshops were uploaded onto Moodle. This redefined and transformed how face-to-face workshops were previously conducted.

Moreover, this is aligned with the notions of the Substitution, Augmentation, Modification, and Redefinition Theory Framework (SAMR) model that framed the study under focus. Thus, using digital platforms for this study allowed lecturers to advance, redefine, and transform their traditional pedagogic strategies within the ambit of the SAMR model (Puentedura, 2014; Terada, 2020). As was evident, academics need to be creative when designing and using digital pedagogy to promote meaningful learning experiences for their students (Sailin and Mahmor, 2018). Furthermore, Cheung and Slavin's (2013) research has shown that using digital devices, tools, and resources for teaching and learning has supported student performance. Bruce (2012) has maintained that using digital devices, tools, and resources when teaching may assist students with challenges and those who thrive when learning. However, although digital tools, devices, and resources are widespread globally, many students and lecturers are still not competent technology users (Jeong, 2017). Also, not all lecturers and students can access digital tools, devices, and resources required to participate equally in a digital pedagogy-based environment. Thus, it is important to research the strengths, challenges, and implications of digital pedagogy for mathematics education.

## Theoretical Framing

The SAMR model allows academics to think about how and why they use technology-based pedagogy to allow technology to support them to evolve pedagogically (Puentedura, 2014). The SAMR model represents "Substitution (S),

Augmentation (A), Modification (M), and Redefinition (R) and offers four ways of how technology can be integrated into pedagogy" (Bouchrika, 2023, p. 2).

Regarding Substitution, technology is used as a substitute for traditional classroom devices. In this way, technology is used to deliver pedagogy with no change to how the pedagogy functions. Regarding Argumentation, technology-based tools, devices, and resources are used to improve teaching, learning, and assessment activities. When looking at Modification, technology-based tools, devices, and resources provide support to bring about a significant change to pedagogy. Lastly, with Redefinition, technology-based tools, resources, and devices are used to entirely transform traditional pedagogy (Terada, 2020). Thus, when Substitution and Augmentation are grouped, they are used as enhancement tools for pedagogy (Daniela, 2021). Modification and Redefinition are tools for transforming pedagogy (Best, 2020). For this study, the SAMR model was used to analyze and reflect on the data generated to interrogate if the digital tools, when integrated within digital pedagogy, enhanced or transformed pedagogy within the education context under study.

## Research Methods and Design
### General background
This chapter reports on a study focusing on postgraduate students' experiences of digital pedagogy for mathematics education. This qualitative study was positioned within an interpretive paradigm. Data were generated from participants at one teacher education institution in South Africa. Participants were purposively selected for ease since the researchers taught or supervised these students. The study incorporated two interactive online workshops and two online discussion forums with participants.

### Participants
Before the study commenced, participants were emailed an informed consent sheet explaining the study's purpose and process. Pseudonyms were used to protect the anonymity and privacy of participants. The participants were made aware that they could leave the study without prejudice. The population for the study was postgraduate mathematics teacher education students. These students were also mathematics schoolteachers. Sixty-nine participants taught or supervised by the researchers were invited to participate. Due to various reasons, approximately 60% ($N=41$) of the participants agreed to participate in the study (25 females and 16 males). Eleven participants (6 females and 5 males) were selected randomly to participate in the pilot study. The remaining 30 participants (19 females and 11 males) participated in the main study.

## Pilot study

Conducting the pilot study increased the dependability and trustworthiness of the workshops and discussion forums for the study. During the pilot study, Internet connections and access to electricity were limited due to load shedding in the different areas of South Africa where the participants participated. As a result, the workshops were sometimes interrupted and took longer than expected. To avoid similar challenges during the main study, the online workshops were held at a convenient time for all participants through negotiation with the participants about Internet connections and access, and by carefully studying the national load-shedding schedules. Also, the workshop and discussion forum questions were reviewed in collaboration with the pilot study participants and other colleagues working in the mathematics education field. This was done to enhance the dependability of the research process and instruments.

## Main study

Thirty participants (19 females and 11 males) participated in the main study. Data were generated via two interactive online workshops and two online discussion forums. Although 30 participants were involved in both interactive workshops, due to various reasons, 22 participants (13 females and 9 males) actively participated in both discussion forums. Pseudonyms were used to protect their identity and guarantee the participants' privacy. The pseudonyms used for this study are shown in Table 5.1.

Table 5.1: Pseudonyms used for the 22 participants who actively participated in the two discussion forums

| Participant # | Pseudonym | Participant # | Pseudonym |
|---|---|---|---|
| 1 | Akhona | 12 | Lionel |
| 2 | Bella | 13 | Lucy |
| 3 | Bongani | 14 | Mandisa |
| 4 | Carol | 15 | Nqobile |
| 5 | Clark | 16 | Olwethu |
| 6 | Deon | 17 | Phumlani |
| 7 | Edwin | 18 | Sandra |
| 8 | Fred | 19 | Sipho |
| 9 | Gugu | 20 | Thabile |
| 10 | Jabu | 21 | Themba |
| 11 | Jane | 22 | Zinhle |

## Online workshops

Online mathematics education workshops (N=2) using Microsoft Teams and Zoom (digital platforms used at the participating university) were conducted with the participants for this study. The researchers facilitated these workshops. Before each workshop, the participants were emailed with relevant material, PowerPoint presentations, and examples of online assessments. The participants were also given information and resources about using digital pedagogy for mathematics education. The researchers shared these resources and information with students online via email, Zoom, Microsoft Teams, and WhatsApp. For this study, the first workshop focused on online mathematics assessments and providing online feedback to mathematics students. The second workshop focused on identifying and resolving student misconceptions in mathematics. At the end of the second interactive workshop, all participants were invited to participate in digital discussion forums (Moodle/Learn 2021 discussion forum and WhatsApp). The discussion forums were designed to explore postgraduate students' experiences of digital pedagogy for mathematics education.

## Online discussion forums

The discussion forums aimed to explore postgraduate students' experiences of digital pedagogy for mathematics education. The discussion forums were conducted via Moodle/Learn 2021 and WhatsApp online chats. Questions or comments were placed on these digital platforms, and participants could respond and comment at their leisure during the week. If clarification were required, the researchers would probe participants' responses further. Each discussion forum began with a few common questions to place the participants at ease. The discussion forum then advanced to specific items focusing on the participants' experiences of digital pedagogy for mathematics education.

While explaining the research process, the participants were asked to reply at least once to each question on both discussion forums. The discussion forums concentrated on the following key questions:

1. What were postgraduate students' experiences of digital pedagogy for mathematics education?
2. What were the strengths of digital pedagogy for mathematics education?
3. What were the challenges of digital pedagogy for mathematics education?
4. Based on postgraduate students' experiences, what were the implications of digital pedagogy for mathematics education?

At the end of each discussion forum, participants were asked to read through their responses and to make revisions or further input if necessary.

## Data analysis

Qualitative data analysis started with the coding of responses. Subsequently, responses were collated, and finally, responses were categorized into major themes. The study's theoretical framing, that is, the SAMR model, was used during the data analysis process. The relationship between the data generated and the notions of SAMR was examined and explored. The qualitative data analysis comprised the following steps: first, open coding was used to ensure the researchers were well acquainted with the data. Second, the data that were related to each other were coded and grouped into themes. Third, the themes were carefully examined to warrant that all codes for each theme revealed a connection. Finally, the relationship between the participants' responses and the SAMR model was explored. Also, member checking was used to ensure the accuracy of the captured responses.

## Ethical considerations

Ethical clearance for this study was acquired from the Research Ethics Department at the participating university. Participants completed informed consent forms prior to joining the study. Pseudonyms were used to ensure participants' privacy and anonymity.

# Findings and discussion
## Learning time is extended when using digital pedagogy

Participants indicated the strengths that digital pedagogy offered them. For example, they had more time to work on their assessment tasks and engage with the material for their online courses. The use of substituting paper copies of module material with online material transformed teaching and learning and enhanced students' learning experiences. This conforms to the notions of the SAMR theoretical framework model. Hence, this was a different experience from what students usually experienced during face-to-face contact teaching and learning. These notions are reflected in excerpts taken from selected discussion forum extracts.

> **Bella:** … more time to look and work with the maths tasks … they were uploaded on Moodle [Learn 2021][3] at the start of the course …
> **Deon:** … it was different … generally we are handed tasks at the start of the lecture when we met lecturers face-to-face … online learning … maths tasks are placed on the Learn platform early …

---

[3] The participating university changed the name of its official online learning platform from Moodle to Learn 2021 during the COVID-19 pandemic.

Gugu: … I had more time to work with the maths content … I could reflect on what I was learning at my own pace … I could ask my friends for help outside lecture times … it worked well for me …

Lionel: … our lecturers gave us extra time to work on maths assessments … we were also given extra catch-up for maths assessments …

Sipho: … all maths lectures were recorded, and we could watch many times if we did not understand something … we had more time … we could also ask for help on WhatsApp and Moodle … we had a better support network than when we worked with regular contact lectures …

Thus, having more time to reflect on tasks and content material benefits students (Chang, 2019). This supports and promotes sustainable learning (Nanjundaswamy et al., 2021). Also, besides enhancing content knowledge of the material being studied, increasing learning time encourages self-confidence and improved study skills and motivates students to commit to learning (Kidron and Lindsay, 2014). Students can use time outside lectures to reflect on and grasp concepts not understood during lecture time. Consequently, lecturers extended the learning and assessment time. They enhanced digital pedagogy by substituting hard copies of assessment tasks and content material with uploaded assessment tasks and content material ahead of the due dates (Daniela, 2021). In this way, lecturers used digital pedagogy to allow technology to support them in evolving pedagogically, thereby augmenting digital pedagogy, which is aligned with the fundamental notions of the SAMR model (Puentedura, 2014).

## The need for preparation workshops for students and lecturers for successful digital pedagogy

The participants indicated that although using technology, digital tools, platforms, and devices was exciting and interesting, it was new to them and their lecturers. As a result, challenges were flagged. Participants were of the notion that more preparation and training workshops for the effective use of digital pedagogy would have been of benefit to their lecturers and themselves. These notions are within the ambit of the SAMR theoretical model. Effective training is needed to allow lecturers the opportunity to redefine and modify their teaching strategies. These sentiments are reflected in the selected discussion forum extracts that follow.

Akhona: … I need more workshops … new and good to learn …

Bongani: … we need time to learn to use the tools … we were just told to attend maths lectures online … there was an online video if we had problems with Zoom, and that was it … but it was new to us … scary at first …

Edwin: … our maths lecturers tried their best … this was also new to them … we tried to help each other … it is important for more training sessions for all of us …

| | |
|---|---|
| **Jabu:** | … I knew how to use Moodle … but I think the new students should have had more training with online learning … some of the online maths tools are a bit complicated … |
| **Lucy:** | … we were just told what to do … it was a shock to us … we always worked with contact maths lectures; this online learning is new to us … the Hovercam was also new for us … used to using the data projector in our maths lectures. |
| **Sandra:** | … I think we all want to do well … the maths lecturers want to help us … we want to succeed … if we could have more training and support on how to use technology and all these new gadgets … it would be better for us … |
| **Thabile:** | … I never used Zoom and Teams … so this was something new and interesting … there was an online video to help us … but I think if we had more training and help, it would be easier … the university has a team we could email if we have problems … but we need more than this … |

While innovative pedagogy piques students' interest and knowledge, exposure and adequate training to use these tools, devices, and platforms are important. Research shows that there are limited training opportunities available for academics, which creates a barrier to the effective use of digital pedagogy (Bescherer, 2020; Ghavifekr et al., 2016; Jeong, 2017; Palaigeorgiou and Grammatikopoulou, 2016). When designing activities while using digital pedagogy, academics need to ensure that students are provided with meaningful learning experiences (Sailin and Mahmor, 2018; Van den Heuvel-Panhuizen and Drijvers, 2020). Thus, academics need to be provided with adequate training to ensure that this outcome is met. With adequate training and preparation, lecturers can decide how best to integrate technology into pedagogy (Bouchrika, 2023). Technology-based tools, devices, and platforms enhance and augment digital pedagogy (Daniela, 2021).

## Active collaborative engagement needs to be encouraged when using digital pedagogy

The participants indicated that there were important implications for the successful use of digital pedagogy. For example, participants mentioned that active collaboration needs to be encouraged to use digital pedagogy effectively. These views align with the SAMR theoretical framework model in that transforming the teaching and learning environment requires modifying traditional teaching and learning strategies. These views are reflected in the selected discussion forum extracts that follow.

| | |
|---|---|
| **Jane:** | … need to work together … we are isolated with COVID-19 … need to support each other during these difficult times … |
| **Nqobile:** | … we need to all help each other … we need more interaction on the maths group chats … the same people talk on the chat … others are silent |

| | |
|---|---|
| | … we don't know if they are ok or if they are having problems … we want to help everyone … |
| **Phumlani:** | … we need to discuss and work together … if we are in class when we see each other, we can discuss and help … in the Zoom maths workshops, we don't know each other … we can't see each other … it's strange … I need to get used to this, but it is hard not discussing or talking to everyone … we need to remember to raise our hands on Zoom … it is very different … |
| **Themba:** | … I am still getting used to working online … we need help from each other … when we are in the class, it is easier to work together … with the Teams group, it is difficult … there is little interaction … we need more interaction … used to working together usually … |

It is important to encourage active, collaborative engagement within an online learning environment. We need to incorporate digital pedagogy as a valuable way to enhance learning and foster notions of collaboration within the digital learning environment. The absence of peer collaboration may have a negative effect on the learning process (Gilbert, 2015).

Moreover, meaningful student learning experiences are supported by active and constructive collaboration within the digital learning environment (Sailin and Mahmor, 2018). Modifying traditional contact-based collaborative discussions and encouraging collaboration via online discussion forums ensures that discussions are significantly transformed with technology. Technology is thereby effectively integrated into pedagogy (Bouchrika, 2023). This integration is essential within the ambit of the SAMR model (Puentedura, 2014).

## The importance of access to digital resources, platforms, and tools for success with digital pedagogy

Participants mentioned another implication for the effective use of digital pedagogy. Participants indicated that for digital pedagogy to be successful, all students and lecturers within the digital learning environment need access to essential digital resources, tools, and platforms. These are important views since, for substitution and augmentation within the ambit of the SAMR theoretical framework model, lecturers and students must have access to digital tools and devices. These ideas are reflected in the selected discussion forum extracts that follow.

| | |
|---|---|
| **Carol:** | … I can't always log in to Moodle or the online workshops because the Internet connection where I live is not always stable … |
| **Clark:** | … for online learning to work, we need all the devices and internet access … I don't have a laptop … I thought I would use the computers at university for this course … |

> **Fred:** … living in a rural area makes it difficult to do online learning … electricity cuts and unstable Internet networks are common in my area …
> **Mandisa:** … with lockdown conditions because of COVID-19, we are not allowed to travel to the Internet shop in my area … I rely on my cell phone for online learning … I don't think it is fair … sometimes I can't use my cell phone to participate in the online workshops …
> **Olwethu:** … I don't have a laptop, and I don't have internet access at home … I need to rely on my cell phone, and the cell network is not good where I live …
> **Zinhle:** … I don't always have data on my phone to join the online maths workshops … but the maths workshops are recorded and uploaded on Moodle … I can look at the recordings when I buy data … but this is not good since I can't ask questions during the workshop …

As is evident, access to digital tools, devices, and platforms is essential for digital pedagogy within the context of the 4IR and COVID-19 (Naidoo, 2020). Moreover, digital platforms, such as Moodle, support students and improve their performance in mathematics (Handayanto et al., 2018). Thus, students' and lecturers' access to technology-based devices, the Internet, and data is essential since access to these crucial resources has significant implications for online collaboration and achievement (Gilbert, 2015; Naidoo & Govender, 2021). Consequently, technology is used to modify or redefine tasks and transform digital pedagogy (Daniela, 2021). Thus, when recordings of the workshops were uploaded onto Moodle, this redefined and transformed how face-to-face workshops were previously conducted. Using digital platforms allowed lecturers to advance, redefine, and transform their traditional pedagogic strategies within the ambit of the SAMR model (Puentedura, 2014; Terada, 2020).

## Conclusion

This study sought to explore postgraduate students' experiences of digital pedagogy for mathematics education. The study was framed within the ambit of the SAMR model, which signifies the use of technology for Substitution (S), Augmentation (A), Modification (M), and Redefinition (R) within a digital education environment. The findings of this study respond to the main and sub-research questions. Participants provided their detailed experiences of digital pedagogy via the discussion forums. Based on these detailed descriptions, as presented in the Findings and Discussion section, the participants discussed the strengths and challenges of digital pedagogy for mathematics education. In addition, the participants signaled key ideas regarding the implications of digital pedagogy for mathematics education.

First, in this study, the strengths of using digital pedagogy for mathematics education were evident when lecturers extended learning time to support students in achieving success within a digital mathematics learning environment. This exhibits an advancement and transformation in pedagogy. Second, another strength that the participants mentioned was that the use of technology enhances pedagogy, and this notion is aligned with the fundamental concepts of the SAMR model. Third, this study shows that active collaborative engagement within an online learning environment to support meaningful learning experiences that promote sustainable learning is valuable. The participants viewed this as a strength of using digital pedagogy. Similarly, the SAMR model supports the idea that technology can modify traditional contact-based collaborative discussions and encourage collaboration via online discussion forums. In this way, online group discussions are significantly transformed with technology.

Fourth, one of the key challenges revealed in this study is the lack of training for students and lecturers to use digital pedagogy effectively. Thus, the findings of this study suggest that if adequate training is provided to both students and lecturers, both can make an informed decision about which digital tool, device, or platform is most suitable for successful digital pedagogy. Thus, technology may be used as a means of substitution or augmentation within the digital educational environment. The SAMR model supports the notion that suitable digital tools, devices, and platforms can enhance pedagogy. Training and preparation workshops would benefit lecturers globally as we contemplate using digital tools, devices, and platforms to enhance and transform mathematics pedagogy. Lastly, it was evident in this study that one of the challenges included inadequate access to digital tools and devices. Access to digital tools, devices, platforms, and Internet data is essential. Access to these tools, devices, and platforms supports lecturers in modifying and redefining tasks within the digital education environment. The SAMR model supports the idea that digital devices, tools, and platforms enhance and transform traditional pedagogy. Thus, our findings highlight key aspects that ought to be considered to ensure the success of digital pedagogy in mathematics education both nationally and internationally. These findings have implications and relevance for developing countries and provide a foundation for encouraging further research possibilities.

Further research possibilities include the need for comparative studies in national and international higher education contexts. In addition, a bigger sample size may be considered, which includes both undergraduate and postgraduate students. This will allow researchers to use quantitative research instruments to generate and analyze data from a larger sample size. Moreover, future research opportunities may provide opportunities for policy revisions and decision-making regarding the integration of digital pedagogy within mathematics higher education.

## Bibliography

Alabdulaziz, M. S. (2021). COVID-19 and the use of digital technology in mathematics education. *Education and Information Technologies* 26(1), 7609–7633. <https://doi.org/10.1007/s10639-021-10602-3>.

Bescherer, C. (2020). Technologies in mathematics education. In A. Tatnall (Ed.), *Encyclopedia of Education and Information Technologies* (pp 1705–1718). Springer. <https://doi.org/10.1007/978-3-030-10576-1_27>.

Best, J. (2020). *The SAMR Model Explained (with 15 practical examples)*. Retrieved February 01, 2023, from <https://www.3plearning.com/blog/connectingsamrmodel/>.

Boholano, H. B. (2017). Smart social networking: 21st century teaching and learning skills. *Research in Pedagogy* 7(1), 21–29. <https://eric.ed.gov/?id=EJ1149146>.

Bouchrika, I. (2023). *How to Use SAMR Model in Designing Instruction (An EdTech Integration Guide)*. Retrieved February 01, 2023, from <https://research.com/education/how-to-use-samr-model-in-designing-instruction>.

Bruce, C. D. (2012). Technology in the mathematics classroom: Harnessing the learning potential of interactive whiteboards. *What Works? Research into Practice* 38(1), 1–4. <https://www.semanticscholar.org/paper/Technology-in-the-Mathematics-Classroom-Harnessing/b43aa939f599c7dd12b269d8fb2b19da26745003>.

Buzzard, C., Crittenden, V. L., Crittenden, W. F., & McCarty, P. (2011). The use of digital technologies in the classroom: A teaching and learning perspective. *Journal of Marketing Education* 33(2), 131–139. <https://doi.org/10.1177/0273475311410845>.

Chang, B. (2019). Reflection in learning. *Online Learning* 23(1), 95–110. <https://doi.org/10.24059/olj.v23i1.1447>.

Cheung, A. C. K., & Slavin, R. E. (2013). The effectiveness of educational technology applications for enhancing mathematics achievement in K-12 classrooms: A meta-analysis. *Educational Research Review* 9(1), 88–113. <https://doi.org/10.1016/j.edurev.2013.01.001>.

Daniela, L. (2021). Pedagogical considerations for technology-enhanced learning. In D. Scaradozzi, L. Guasti, M. Di Stasio, B. Miotti, A.

Monteriù, & P. Blikstein (Eds.), *Makers at School, Educational Robotics and Innovative Learning Environments. Lecture Notes in Networks and Systems* (pp. 57–64). Springer. <https://doi.org/10.1007/978-3-030-77040-2_8>.

Drijvers, P. (2013). Digital technology in mathematics education: Why it works (or doesn't). *PNA* 8(1), 1–20. <https://doi.org/10.1007/978-3-319-17187-6_8>.

Ghavifekr, S., Kunjappan, T., Ramasamy, L., & Anthony, A. (2016). Teaching and learning with ICT tools: Issues and challenges from teachers' perceptions. *Malaysian Online Journal of Educational Technology* 4(2), 38–57. <https://eric.ed.gov/?id=EJ1096028>.

Gilbert, B. (2015). *Online Learning Revealing the Benefits and Challenges*. St. John Fisher College. New York: Rochester. <https://fisherpub.sjfc.edu/education_ETD_masters/303/>.

Goertz, P. (2015). 10 Signs of a 21st Century Classroom. In *George Lucas Educational Foundation*. <https://www.edutopia.org/discussion/10-signs-21st-century-classroom>.

Handayanto, A., Supandi, S., & Ariyanto, L. (2018). Teaching using Moodle in mathematics education. *Journal of Physics: Conference Series 1013*(012128), 1–4. <https://doi.org/10.1088/1742-6596/1013/1/012128>.

Jeong, K.O. (2017). Preparing EFL student teachers with new technologies in the Korean context. *Computer Assisted Language Learning 30*(6), 488–509. <https://doi.org/10.1080/09588221.2017.1321554>.

Kidron, Y., & Lindsay, J. (2014). *The Effects of Increased Learning Time on Student Academic and Nonacademic Outcomes: Findings from a Meta-analytic Review*. R.E.L.A. National Center for Education Evaluation and Regional Assistance. <https://files.eric.ed.gov/fulltext/ED545233.pdf>.

Moloi, T., & Mhlanga, D. (2021). Key features of the Fourth Industrial Revolution in South Africa's basic education system. *Journal of Management Information and Decision Sciences 24*(5), 1–20. <https://www.etdpseta.org.za/education/sites/default/files/2021-09/Key-features-of-the-fourth-industrial-revolution-in-South-Africas-basic-education-system.pdf>.

Montrieux, H., Vanderlinde, R., Schellens, T., & De Marez, L. (2015). Teaching and learning with mobile technology: A qualitative explorative study about the introduction of tablet devices in secondary education. *PLOS ONE 10*(12), 1–11. <https://doi.org/10.1371/journal.pone.0144008>.

Murgatrotd, S. (2020). *COVID-19 and Online Learning*. Retrieved on April 14, 2020. <https://www.researchgate.net/publication/339784057_COVID-19_and_Online_Learning>.

Naidoo, J. (2020). Postgraduate mathematics education students' experiences of using digital platforms for learning within the COVID-19 pandemic era. *Pythagoras, 41*(1), 1–11. <https://doi:10.4102/pythagoras.v41i1.568>.

Naidoo, J., & Govender, R. (2021). Postgraduate mathematics education students' perceptions of technology-based tools and resources: Exploring the influences of connectivism and the three worlds of mathematics. *Universal Journal of Educational Research, 9*(6), 1214–1223. <https://doi.org/10.13189/ujer.2021.090610>.

Nanjundaswamy, C., Baskaran, S., & Leela, M. H. (2021). Digital pedagogy for sustainable learning. *Shanlax International Journal of Education 9*(3), 179–185. <https://files.eric.ed.gov/fulltext/EJ1300885.pdf>.

Palaigeorgiou, G., & Grammatikopoulou, A. (2016). Benefits, barriers and prerequisites for Web 2.0 learning activities in the classroom: The view of Greek pioneer teachers. *Interactive Technology and Smart Education 13*(1), 2–18. <http://dx.doi.org/10.1108/ITSE-09-2015-002>.

Peachey, N. (2017). *Digital Tools for Teachers*. PeacheyPublications.Com. <https://peacheypublications.com/books/digital-tools-for-teachers>.

Pope, S., & Mayorga, P. (2019). *Enriching Mathematics in the Primary Curriculum. Learning Matters.* <https://doi.org/10.4135/9781526489715>.

Puentedura, R. (2014). *SAMR and Bloom's Taxonomy: Assembling the Puzzle.* <https://www.commonsense.org/education/blog/samr-and-blooms-taxonomy-assembling-the-puzzle/>.

Pyper, J. S. (2017). Learning about ourselves: A review of the mathematics teacher in the digital era. *Canadian Journal of Science, Mathematics and Technology Education 17*(3), 234–242. <https://doi.org/10.1080/14926156.2017.1297509>.

Rasanen, P., Laurillard, D., Käser, T., & von Aster, M. (2019). Perspectives to technology-enhanced learning and teaching in mathematical learning difficulties. In A. Fritz, V. Haase, & P. Rasanen (Eds.), *International Handbook of Mathematical Learning Difficulties* (pp. 733–754). Springer. <https://doi.org/10.1007/978-3-319-97148-3>.

Sailin, S. N., & Mahmor, N. A. (2018). Improving student teachers' digital pedagogy through meaningful learning activities. *Malaysian Journal of Learning and Instruction* 15(2), 143–173. <https://files.eric.ed.gov/fulltext/EJ1201692.pdf>.

Terada, Y. (2020). A powerful model for understanding good tech integration. *Edutopia*. February 01, 2023. Retrieved from <https://www.edutopia.org/article/powerful-model-understanding-good-tech-integration/>.

Van den Heuvel-Panhuizen, M., & Drijvers, P. (2020). Realistic mathematics education. In S. Lerman (Ed.), *Encyclopedia of Mathematics Education* (pp. 521–525). Springer. <https://doi.org/10.1007/978-3-030-15789-0_170>.

CHAPTER 6

# Teachers' Experiences of Blended Classrooms in Mathematics, Science, and Technology Academic Hubs

*Zingiswa Jojo[1] and Puleng Motseki[2]*
[1]Rhodes University
[2]University of South Africa

**ABSTRACT**
This chapter explores teachers' experiences of blended classrooms in science, technology, engineering, and mathematics (STEM) academic hubs in Mpumalanga province in South Africa before, during, and after the COVID-19 pandemic era. Espousal of the ideals of the Technological Pedagogical Content Knowledge framework is considered a crucial theoretical approach for researchers to understand how teachers use technology in their classrooms to focus on the kinds of knowledge that teachers require to use technology effectively in the classroom. From a qualitative research approach, a case study research design was followed using a purposive sample of four coordinating lead teachers who participated in this study. Data were collected through semi-structured interviews. Thematic analysis was used to identify, analyze, and report the findings. The five themes that emerged from the analysis are collaboration on information and communications technology (ICT)-related matters, technical support, availability of e-learning material, availability of ICT infrastructure, and pedagogical use of ICT resources as innovative instructional approaches used by mathematics teachers in complementing face-to-face teaching with online mathematics lessons to enhance learners' understanding of mathematics concepts during the COVID-19 era. The chapter ends with the highlights of the implications of the study to the wider STEM education community.

*Keywords:* Asynchronous learning, blended learning, e-learning, STEM, synchronous

## Introduction

The COVID-19 pandemic disrupted education in over 150 countries worldwide and affected over 1.6 billion learners (Vincent-Lancrin et al., 2022). The World Health Organisation (WHO) declared COVID-19 a global pandemic on 11 March 2020. In response to COVID-19, several countries, including South Africa, imposed strict social distancing measures, including the lockdown. The implementation of strict social distancing policies and lockdowns had a dreadful impact on curriculum implementation, learners, and teachers in schools. For instance, soon after the schools' closure, UNESCO (2020) predicted that 23.8 million learners would not return to school after a 3-month closure.

Moreover, studies focusing on Sub-Saharan Africa predicted up to 3 years of lost learning (Angrist et al., 2021). In South Africa, the government restricted community mobilization to prevent the rapid spread of the COVID-19 virus and promoted the idea of hybrid working and learning. Schools and institutions of higher learning in South Africa were temporarily closed from March 16, 2020. Within the South African context, there is a dearth of research that explores teachers' experiences of blended teaching, especially in the science, technology, engineering, and mathematics (STEM) academic hubs. In addition, the STEM researchers have not extended the research to explore the nature of teaching and learning in rural areas of Mpumalanga province of South Africa to offer alternative forms of knowledge to the urbanized research knowledge.

To save the academic year 2020, remote teaching and learning were implemented using information and communication technologies (ICT). Electronic learning (e-learning) was considered the best approach to continue teaching and learning during the pandemic. Nicholson (2007) defines e-learning as a form of learning that uses the affordances of the Internet to deliver customized, often interactive learning materials and programs to diverse local and distant communities of practice. However, the implementation of e-learning is not always successful. During the COVID-19 pandemic, schools with limited experience and unprepared e-learning resources experienced challenges, especially with teachers with no relevant pedagogical knowledge of how to use online applications (Zaharah et al., 2020).

Several studies (Coetzee et al., 2021; Gumede and Badriparsad, 2022; Gamede et al., 2022) on the integration of e-learning in mathematics teaching were conducted in South Africa during the COVID-19 pandemic in higher education contexts. For instance, Pule and Ngoveni (2024) conducted a quantitative study on second-year mathematics pre-service teachers at a rural university who had to adapt to online learning due to the COVID-19 pandemic. The findings reveal that the pandemic has had an even more significant impact on vulnerable populations residing in rural areas due to their limited access to online learning. This implies that higher education institutions in socio-economically challenged communities that lack resources contribute to education disparities compared to their urban counterparts. The study by Pule and Ngoveni (2024) also reveals challenges experienced by teachers in blended mathematics classrooms. These relate to teachers' struggle to adjust to mathematical, scientific, and technological progress that demands creativity and flexibility, the lack of which had a negative impact on learners' performance and behavior. This highlights the importance of exploring secondary school teachers' experiences integrating e-learning into traditional mathematics classrooms during the COVID-19 pandemic.

The STEM Academy was established to increase the number of learners taking STEM subjects at the high school level. Its aim was also to improve the quality of success in mathematics and physical sciences, assist with the training of mathematics, science, and technology (MST) teachers, enhance and develop teachers' competency levels in the effective teaching of those subjects, strengthen the integration of ICT in teaching MST subjects and develop strategies that enhance learner performance (Netshakhuma, 2020). During the past 3 year (2021–2023), the province has seen an increase in the number of learners taking mathematics, sciences, and technology through Grades 10–12 and has registered a high throughput rate.

Table 6.1: Summary of Grade 9 mathematics performance in sub-Hub A

| Sub-Hub | Year | No. wrote | (0-29%) | (30-39%) | (40-49%) | (50-59%) | (60-69%) | (70-79%) | (80-100%) | Overall %Pass | % Pass at 50% | Average % |
|---|---|---|---|---|---|---|---|---|---|---|---|---|
| A | 2020 | 178 | 178 | 0 | 0 | 0 | 0 | 0 | 0 | 0 | 0 | 0 |
|   |      |     | 100 | 0 | 0 | 0 | 0 | 0 | 0 |   |   |   |
|   | 2021 | 162 | 49 | 44 | 24 | 2 | 1 | 0 | 0 | 32,72 | 5.56 | 30,09 |
|   |      |     | 30,25% | 27,16% | 14,8% | 1,23% | 0,62% | 0 | 0 |   |   |   |
| B | 2019 | 168 | 0 | 0 | 76 | 52 | 26 | 13 | 1 | 100 | 54,76 | 52,1 |
|   |      |     | 0 | 0 | 45,2 | 30,9 | 15,4 | 7,7 | .6 | 100 |   |   |
|   | 2018 | 179 | 0 | 0 | 146 | 15 | 11 | 5 | 2 | 100 | 18,32 | 45,99 |
|   |      |     | 0 | 0 | 81,56 | 8,38 | 6,15 | 2,79 | 1 |   |   |   |

Despite Mpumalanga province having recorded an increase in the number of learners taking MST subjects from Grades 10–12 with evidence of a high throughput rate, Grade 9 performance in mathematics in one of the sub-hubs continued to decline. Table 6.1 summarizes red flags in STEM subject performance in Grade 9, especially in mathematics in 2020 and 2021. Few learners passed mathematics at 50% and above. In addition, in this sub-hub in 2020 and 2021, no learners passed mathematics with distinction in Grade 9. This may be due to several factors, including the necessity of using a blended learning approach at the time. Although other factors played out and caused low scores in mathematics performance, the COVID-19 impact on those results cannot be ruled out. Blended learning is an innovative concept that embraces collaborative, constructive and computer-assisted learning. Since its modes of teaching and learning draw on technology-mediated instruction, where all participants in the learning process are separated by distance most of the time, this might have rendered challenges to teachers of this hub due to a lack of training. The current study aimed to explore teachers' experiences of blended mathematics classrooms in STEM academic hubs during the COVID-19 pandemic.

To achieve this aim, this study aspired to answer the following research questions:

1. What affordances do mathematics teachers perceive as significant to blended learning during the pre- and post-COVID-19 pandemic?
2. What challenges do mathematics teachers perceive as significant to blended learning during the pre- and post-COVID-19 pandemic?

## Literature Review

The literature presented in this chapter discusses (i) the Technological Pedagogical Content Knowledge framework (TPACK), (ii) Affordances of e-learning integration, and (iii) e-learning integration barriers.

### Technological pedagogical content knowledge framework (TPACK)

This study was underpinned by the TPACK framework by Mishra and Koehler (2008). One way in which researchers try to understand how teachers integrate technology in their classrooms is to focus on the kinds of knowledge that teachers require to use technology effectively. Shulman (1986) suggests that for blended teaching to be effective, teachers require a special type of knowledge known as pedagogical content knowledge (PCK). This form of knowledge represents "the blending of content and pedagogy into an understanding of how particular topics, problems, or issues are organized, represented and adapted to the diverse interests and abilities of learners, and presented for instruction" (Shulman, 1986, p. 8). Mishra and Koehler (2008) proposed that to teach a particular subject matter, teachers require not only knowledge of the content but must develop appropriate instructional strategies and skills that are appropriate for learners. Shinas et al. (2015) advise that TPACK development is possible when teachers are trained and prepared in a comprehensive program that enhances technology-integrated lesson planning. The TPACK framework suggests that educators should have adequate technological, pedagogical, and content knowledge to successfully enhance student learning, no matter which environment they are working in. Mishra and Koehler's (2008) development of the TPACK framework extended Shulman's (1986) PCK theory to consider the role that technology can play in providing teachers with an opportunity to engage learners with mathematics content that would otherwise be less accessible if presented in a textbook, for example, exploring the properties of three-dimensional geometry objects.

The three main forms of knowledge that are central to the foundation of TPACK are:

a. Content Knowledge (CK) refers to any subject matter knowledge the teacher should have for teaching.
b. Pedagogical Knowledge (PK) refers to the teacher's knowledge about various instructional practices, strategies, and methods to promote learning.
c. Technology Knowledge (TK) refers to the knowledge the teacher should have regarding traditional and new technologies that can be included in the learning process.

The following four forms of knowledge in the TPACK framework address how the three bodies of knowledge interact, constrain, and afford each other.

a. Technological Content Knowledge (TCK) refers to the knowledge of the reciprocal relationship between technology and content. Disciplinary knowledge is often defined and constrained by technologies and their representational and functional capabilities.
b. Pedagogical Content Knowledge (PCK) refers to Shulman's (1986) idea of "an understanding of how particular topics, problems, or issues are organized, represented, and adapted to the diverse interests and abilities of learners, and presented for instruction" (p. 8).
c. Technological Pedagogical Knowledge (TPK) refers to the understanding of how technology can constrain and afford specific pedagogical practices and
d. TPACK refers to knowledge about complex relations between technology, pedagogy, and content that enable teachers to develop appropriate and context-specific teaching strategies. Figure 6.1 depicts the three primary knowledge forms central to the TPACK framework.

The TPACK framework proposes that teachers are required to understand the forms of knowledge above to orchestrate and manage technology, pedagogy, and content in teaching (Sointu et al., 2017). Moreover, McGarr and McDonagh (2021) emphasize that the main goal of TPACK is to educate teachers on how to effectively use technology in a variety of educational settings, such as face-to-face instruction, blended learning, online learning, and homeschooling, as well as other types of education. This theory was found relevant in this study since the blended mode of instruction in STEM schools relies mostly on technology tools.

The integration of e-learning in teaching and learning is a complex phenomenon. Thus, many teachers may either use the numerous opportunities available with e-learning or encounter difficulties with it. The opportunities are affordances (Chatterjee et al., 2020), while the difficulties are regarded as barriers

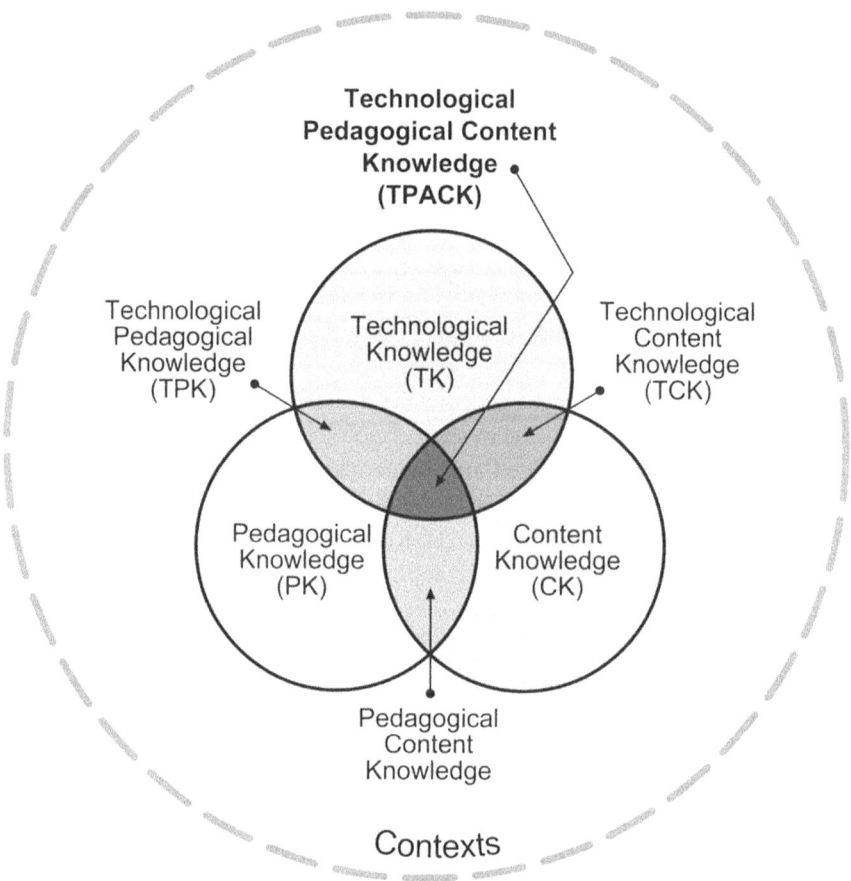

Figure 6.1: The technological pedagogical content knowledge (TPACK) framework.

(Schoepp, 2005). The Merriam-Webster Online dictionary defines affordance as the quality or property of an object that defines its possible uses or clarifies how it can or should be used. In contrast, a barrier is an obstacle preventing movement or access. In this chapter, Chatterjee et al.'s (2020) definition of affordance and Schoepp's (2005) definition of the barrier were adopted. The literature review section focuses on the affordances of e-learning integration and e-learning integration barriers. The rationale for focusing on the two is to understand the lead teachers' experiences of technology as an enabler for effective teaching and learning and the barriers associated with integrating technology in teaching and learning mathematics.

## Affordances of e-learning integration

The quality of learning during the COVID-19 era was primarily determined by integrating technology, information, and communications in mathematics classrooms (Putra et al., 2021). Mishra and Koehler (2008) suggest that technology can help improve instructional processes for the better. However, Clark (2010) cautions that educators are not solely responsible for the learning process, other than the fact that technology integration helps build effective digital relationships with learners within school communities while continuously striving to reflect on and improve instructional practices.

Technology affordances in mathematics classrooms can be seen from the teacher's skills in developing and implementing the e-learning material. Sun et al. (2018) claim that applying e-learning instructional frameworks in mathematics classrooms is fundamental to achieving learning outcomes. Relevant, applicable instructional frameworks provide either synchronous or asynchronous learning. A synchronous learning framework offers flexibility in digital learning without space and time constraints (Putra et al., 2021). This implies instruction and learning that occur simultaneously but not in the same place. It is made possible by creating virtual platforms like video-conferencing discussion groups where teachers and learners interact in "real-time to have maximum time and opportunities to learn" (Zydney et al., 2020). Jethro et al. (2012) regard e-learning as a synchronous environment for effective and efficient learning to improve mathematical knowledge acquisition and learner performance. This learning model is based on the constructivist approach to learning, aimed at motivating learners to solve challenging assignments in problem solving and high-order thinking skills, and helps learners to be actively involved during the learning process. They could be involved in continuous reflection, production, synthesis, and instructional process analysis. This is impossible because learning is a natural social act that occurs through talking, attempting to solve problems, and seeking to understand the world.

Another learning framework widely used by mathematics teachers is the asynchronous learning framework. Asynchronous learning involves using pre-recorded lessons and audio, video, and text files to ensure learners follow up by watching and listening to pre-recorded learning material (Zydney et al., 2020). Thus, the combination of synchronous and asynchronous learning is the trend that was newly implemented in the learning process at the dawn of COVID-19.

However, there are problems with implementing the blended learning approach in mathematics classrooms. Cui and Zheng (2018) caution that in applying the blended learning approach, some teachers use online learning only as a complement to traditional learning and fail to change the teaching process in traditional classrooms. Moreover, this exercise deprives learners of teacher feedback and guidance because of a lack of effective collaboration management skills (Cui and Zheng, 2018). Although there are challenges in implementing blended learning in mathematics, Helsa et al. (2021) acknowledge the benefits of using blended learning: (i) to improve mathematical thinking skills, develop good perceptions, communication skills and learning outcomes, (ii) to increase self-regulation, thinking/problem-solving skills, student participation, together with computational thinking skills, and critical thinking skills, and (iii) to simplify the assessment process.

### e-Learning integration barriers

Various researchers have put forward several types of e-learning barriers. For instance, Ertmer (1999) categorized the barriers into first-order and second-order. Ertmer (1999) states that first-order barriers include hardware, accessibility, and technical support, whereas second-order barriers include pedagogy and beliefs. Other researchers (e.g. Assareh and Bidokht, 2011) refer to e-learning barriers that affect learners, teachers, curriculum, and schools. Barriers related to learners are not limited to financial problems, isolation from peers, inadequate e-learning skills and experience, affection, and social domain. Still, they can also affect learners' performance in mathematics. Those related to teachers include knowledge inadequacies and assessment.

Regarding the curriculum, e-learning barriers include ambiguity, quality, resources, instructional process, and evaluation. Lastly, barriers concerning schools include organizational and structural factors. Various mentioned barriers were prevalent in some of the STEM hubs because some teachers lacked knowledge of assessment and content presentation skills.

A study by Hadija and Shalawati (2017) reported that teachers' lack of time to prepare lessons using technology was a major challenge they experienced. According to Bingimlas (2009), other significant limitations were a lack of technological professional development, inadequate physical resources, limited access to technology, and a lack of technical support, competence, and confidence. However, in this study, the main STEM hub specialists provided technical support and access to technology through smartboards installed in the STEM schools, together with different lessons presented and broadcast from the main hub to those schools. STEM schoolteachers who participated in this study were trained to use smartboards and prepare lessons using their laptops.

## Research Methods

South Africa is made up of nine provinces. This study was conducted in one province, Mpumalanga. Currently, 101 secondary schools in the Mpumalanga province are linked to the STEM Academy central hub, with one satellite school in each of the four regions identified as a sub-hub. Each sub-hub is further connected to a minimum of 24 schools within each region in Mpumalanga. To ensure that support is provided to learners before entering secondary school, each of the 101 secondary schools identified and worked with their feeder primary schools, adding up to 492 feeder schools across that province. Subject lead teachers for mathematics (General Education and Training [GET] and Further Education and Training [FET]), natural/physical sciences, technology (GET), accounting, and life sciences were appointed to support teachers from the STEM academy. The sub-hubs have ICT resources like smartboards to support teaching and learning.

Consequently, from 2019 to 2022, the province has seen an increase in the number of learners taking mathematics, sciences, and technology through Grades 10–12 and has registered a high throughput rate. Our context analysis was limited to four mathematics lead teachers with specific reference to Grade 9 performance in one of the sub-hubs. In a qualitative research approach, researchers perceive the world as complex, dynamic, interdependent, nuanced, unpredictable, and understood through stories. They distrust generalizations and are most comfortable immersed in the details of a specific time and place (Patton, 2015, p. 13). As a result, we chose the qualitative approach as an appropriate approach in using words and texts to facilitate understanding of the social phenomenon (Creswell, 2014).

## Data collection

From an interpretive paradigm of a multiple holistic case study that followed a qualitative approach, a purposive sample of four coordinating mathematics lead teachers was selected as participants in this study (Creswell, 2014). An interpretive approach was used to detail the experiences of four mathematics lead teachers coordinating mathematics lessons from the STEM hubs. Semi-structured interviews (Creswell, 2014) were used as this study's primary data collection technique to offer in-depth narratives with participants. Data for this study were collected in two phases: April 19–20, 2022 and October 10–13, 2022. The two phases allowed analytic depth by enabling the exploration and probing of issues identified in early semi-structured interviews. During the field conversations and structured interviews, the researchers adopted a maximum variability approach to access as many different points of perception as possible on the issues of interest. Participants were selected in consultation with STEM leaders as partners to represent a range of different seniority levels before being approached by authors. The process of data collection was guided by the principle of information power (Malterud et al., 2016), which shifts the focus from the number of participants in the sample to the power of information that the sample holds. The interviews focused on participants' experiences with the opportunities and barriers experienced in blended mathematics classrooms in STEM hubs during COVID-19. Interviews lasted between 60 and 80 min daily for 5 day based on the two-phase hub visits. Participants provided recorded consent to participate in the study. Documents that were shared on site visits were also included as data.

## Data analysis

The ethnographic semi-structured interview data were captured as field notes and audio-recorded reflections at the end of each day's interviews. Debriefing sessions were conducted to enable detailed synthesis, and the authors took notes. The recorded interview data were transcribed verbatim, anonymized, and included in the analysis. Transcripts were coded using Saldana's (2014) open coding. Key themes were identified by evaluating repeated patterns of codes that emerged from the data. This enabled us to develop an analytic framework that was continually updated during the analysis process. Because data collection and analysis were conducted concurrently, we were able to discuss preliminary findings with participants and used their feedback to evaluate our emerging analysis against their experiences. The analysis process was enhanced by drawing from the literature review to make sense of relevant concepts and move from descriptions to the interpretation of the data. The inductive data analysis approach was used as the analytical framework underpinning this study. Using the inductive analysis approach, we sought to understand lead teachers' narrations on the affordances and barriers of integrating technology in Grade 9 mathematics classrooms. A qualitative approach to research sheds light on the STEM hubs' complex cultural and institutional settings (Birkinshaw et al., 2011). Thus, qualitative research is more likely to provide meaningful contexts and clarity to the research questions and concepts within the STEM social settings rather than providing the hypothesis and testing of variables.

# Findings

The findings of this study were summarized into four main themes that emerged during the analysis of semi-structured interview responses on mathematics teachers' lived experiences in blended classrooms in STEM hubs.

## Collaboration on ICT-related matters

The defining feature of the STEM hub was its collaboration on ICT-related matters. This kind of collaboration cannot be disaggregated into the sum of individual competencies but conforms to developing a sense of interdependency and understanding while using a collective ICT knowledge base. Mathematics teachers were asked to explain how they engage in collaborative efforts in sharing knowledge and solving problems on ICT-related barriers. The STEM hubs supported the development of social interactions by allowing teachers to learn together and understand each other's roles and responsibilities, including the roles of the lead teachers.

"*The STEM hubs encouraged teachers to work in groups to share knowledge and solve ICT-related problems.*" She said that part of it was because they all teach Mathematics as a subject, with each teacher being required to know their areas of responsibility, even those not allocated to them.

> *We also have an online Dashboard connected to the Department of Basic Education. Once you punch in the learner ID number on the dashboard, you can retrieve the learner profile [academic information], which greatly helps us plan lessons. (Lead teacher A)*

She further noted,

> *In the Hubs, no teachers 'own' the learners or/and ICT resources. As such, it makes it easier when one teacher is absent. The caretaker teacher knows how far the teacher and learners are with the syllabus and takes over to continue with them until the designated teacher returns. Part of it is the frequency and the way they collaborate within a hub and with other hubs; they are all indoctrinated with similar beliefs. (Lead Teacher A, October 10, 2022)*

The fact that collaboration on ICT matters is an important feature of STEM suggests that teachers engage in developmental processes that address ICT integration, classroom implementation policy, consideration of teacher-to-learner ratios, and basic troubleshooting skills. The automated Dashboard connected to the Department of Education provides the hubs with essential baseline data on each learner profile in the hubs. The data are discussed routinely during collaborations to identify areas of concern that require inclusion in the instructional process. Although teachers believe integrating technology into the classroom is essential to enhance mathematics education, they reported that one of the main challenges was learner absenteeism and, in some cases, lack of parental involvement. This was made clear when Lead Teacher D stated:

> *When you phone parents to come to the meeting, a parent will tell you that they do not have the time because they must work. Moreover, when required to be accompanied by parents to disciplinary meetings, these learners go out and 'buy' parents for a fee to represent them. (Lead Teacher D, October 10, 2022)*

## Technical support

STEM is characterized by an unrelenting insistence on the technical competence of the technical support team to enable mathematics teachers to perform their teaching duties with high proficiency standards. The effects were visible in teachers' confidence, readiness, and competence when responding to technical issues.

> We have a team of technicians based at the head office. The team responds to technical queries such as hardware failure and software malfunction for all the sub-hubs. During the live teaching and learning sessions, all technicians can connect all the sub-hubs concurrently. All the software installations, updates, and anti-virus programs are done on a central server at the head office. As lead teachers, we also have basic technical support skills such as troubleshooting as replenishing a finished toner, long pressing the Esc key on the keyboard for a laptop/desktop that froze and using the network cable to connect a laptop/desktop to a network printer. (Lead Teacher B, October 10, 2022)

Efforts went into socializing junior mathematics teachers into the ethos of collective endeavor and mutual respect across levels of seniority. For example, newly appointed mathematics teachers were provided laptops with all the e-learning content installed and assigned to a lead teacher upon joining the STEM school.

## Availability of e-learning material

STEM is committed to providing e-learning materials to teachers and learners. This became evident from the statement given by Lead Teacher A when she said:

> Each learner has a tablet with mathematics learning material and pre-loaded question papers. In addition to the tablets, each learner is provided with a mathematics textbook that they use to learn. Learners are also provided with monthly one-gig data. Learners and teachers sometimes engage in online mathematics discussions using Google Meet. Learners work on mathematics worksheets and upload them on Google Classroom. In addition to live classroom discussions and pre-loaded learning material, we record 60 lessons for Mathematics GET and Mathematics FET. To some extent, the data is little, considering the work we do with them. Take yourself as an individual; what can you do with one gig of data? This is the same with the teachers. Each teacher is provided with a laptop with learning material pre-loaded. At the end of each year, learners know that they must return the tablets and textbooks so that the next group of learners will use them.

## Availability of ICT infrastructure

Critical to STEM hubs is the ability to provide the required tools of the trade for teachers and learners. This is evident from Lead Teacher D's statement:

> We have a mathematics laboratory with 40 desktops, one smartboard, and an overhead projector. Desktops. Each teacher connects to the smartboard using their laptop. In consultation with the HoD [Head of Department], the lead teacher sits together to develop a schedule/timetable regarding access to the mathematics lab. Currently, more time is allotted to the Grade 12 learners mainly because they will be sitting for their mid-year assessments soon and exiting the schooling system shortly. We communicate with the director at head office quarterly regarding the maintenance and updates of the software packages we use at the technical and academic

> levels. The head office conducts school support visits to monitor the use of ICT resources. The main challenge, as you can see in this venue that we have convened, was a fully equipped computer lab, including the smartboard, the wiring, and the desks, but the surrounding community stole all the computers. It feels increasingly like a battle to keep our computer lab safe. We do not have proper fencing. So far, we are trying to install solid doors and hire security guards to monitor the labs. We use the labs to connect to other hubs to conduct live classes. (Lead Teacher D, April 2022)

The hubs provide high-quality and available ICT infrastructure; however, the school's lack of proper security challenged it.

## Pedagogical use of ICT resources

Through mentorship, situated learning, and role modelling, lead teachers attempt to socialize other teachers into 'how things are done here'. HoDs and lead teachers constantly mobilized the image of STEM as high-performing, disciplined, and hardworking, with expectations that teachers would live up to shared standards.

> We hold meetings within sub-hubs where lead teachers identify training needs at the school level. Each teacher identifies his/her areas of weakness where the pedagogical use of ICT is deemed essential. For instance, several trainings were conducted with teachers on using a Smartboard during teaching to enhance their digital skills. We then organise training sessions based on the needs analysis. Upon return, each lead teacher trains the teachers within the sub-hubs. We hold quarterly regional meetings online or physically, where all teachers in all sub-hubs come together to take stock and sit down and discuss approaches we use to incorporate ICT in classrooms. (Lead Teacher A, April 19, 2022)

Training days provided a vital function of reflecting on areas of strength and concerns regarding ICT pedagogy. In some cases, sharing beliefs regarding ICT integration appeared to be critical in helping to reinforce values of high standards. On an individual level, teachers receive training on ICT skills and ICT integration in mathematics classrooms. In addition, the lead teachers provide additional pedagogical training on the use of ICT resources in the classrooms. When teachers were asked if they were aware of the TPACK framework, one lead teacher said, *"Well, I do not know what the TPACK framework is. However, we get training on using ICT resources in the mathematics classroom."* (Lead Teacher A, April 19, 2022).

## Discussion

Using a positive deviance approach (Lawton et al., 2014), the discussions in this chapter focus on exploring teachers' experiences in blended mathematics classrooms in STEM hubs. The COVID-19 pandemic disrupted traditional teaching and learning in physical classroom settings. As a result, there was a need to transform how STEM teachers present mathematics lessons to learners so that the instructional process is effective. Thus, technology has become increasingly relevant in classrooms; as a result, teachers ought to possess the professional knowledge and skills necessary for effective integration of technology in the class (Sointu et al., 2017).

The TPACK framework assists teachers with professional development knowledge to increase learners' understanding of complex concepts and encourage collaboration. Teachers are, therefore, encouraged to use some form of technological tools during their lessons. In addition, instructional processes with the integration of technology are interactive, inspiring, fun, and challenging, motivating learners to actively participate and providing sufficient space for initiative, creativity, and independence in accordance with learners' talents, interests, and physical and psychological development (Zhao et al., 2016).

The TPACK framework guided the research in data collection and prescribed how it should be analyzed. Using STEM hubs as a case study, four main themes emerged from the data analysis based on answering the research questions. One of the themes that emerged was the collaboration on ICT-related matters. Collaboration between teachers within the same sub-hub, sharing knowledge as a team, and their ability to work with other teachers from other sub-hubs were the defining features of these hubs. This involved the ability to co-teach and discuss experiences related to ICT. Strong social ties and mutual investment among mathematics teachers created a supportive working environment. They generated rich tacit and relational knowledge, including individuals knowing each other's ICT strengths and weaknesses and being able to stand in for one another when another teacher was absent, all of which buttressed confidence in each other's skills. Teachers were not only in charge of the learning process. However, they were able to build effective and collaborative relationships in the STEM community while continuing to reflect and improve learning practices on an ongoing basis (Clark, 2010).

In contrast to the literature (Bingimlas, 2009), the theme of the availability of e-learning material emerged as another defining feature in the STEM hubs. The success in the instructional process was evident from the teachers' abilities to design and use e-learning to achieve the learning objectives (Sun et al., 2018). The flexibility of integrating live-streamed lessons, video-recorded material, and printed additional enrichment learning material complemented the traditional (textbook) teaching approach and created an optimal learning process for both teachers and learners (Tseng et al., 2013). This is because teachers and learners learned the technology and integrated the e-learning material using appropriate instructional steps. Indumathy et al. (2020) note that ICT is useful in education, digital literacy and developing all kinds of resources; infrastructure development;

logistics management; healthcare; livelihood generation and empowerment of masses; e-governance, administration and finance; specialized business and industrial uses; agricultural uses; research and development; economic growth; and poverty alleviation. The instructional process may be strengthened by adapting easily by applying learning models (Chen et al., 2020). Instructional design that combines synchronous and asynchronous models helps learners construct cognitive abilities (Liang et al., 2020). Integrating the asynchronous and synchronous learning frameworks during the instructional process makes it easier for teachers and learners to determine both place and time for the learning process (Wahyudi et al., 2020). Literature by Quadri et al. (2017) identified limited time to develop e-learning material as the most significant barrier that hindered the implementation of e-learning. However, in this study, participants reported the availability of digital educational content, including the material available through the World Wide Web.

The pedagogical use of ICT resources emerged as one of the themes and significant features of STEM hubs. When lead teachers were asked to comment on the pedagogical use of ICT resources, teachers indicated that the availability of ICT infrastructure enables teachers to use ICT resources in their mathematics classrooms. The provision of computer laboratories, the internet, and smartboards made it easier for teachers to record lessons on poorly performed topics. A teacher training program on ICT skills and integration in mathematics lessons was provided. Pedagogical ICT skills were necessary to help teachers' structure and organize their learning processes. This suggested that teachers received sufficient and timely pedagogical and technical support. Supplying enhanced technology tools to schools, along with improved connectivity, will promote teachers' success in delivering lessons virtually to all learners and may positively affect the learners' performance in mathematics. This finding contradicts the study by Bingimlas (2009), who reports limited access to technology and lack of technical support, competence, and confidence as factors that hinder the successful integration of technology in the classrooms.

Participants knew the importance of technical support in integrating ICT into mathematics classes. The lead teachers responded to questions about the availability and quality of technical support, and their responses revealed adequate and frequent hardware and software updates. They indicated that STEM leadership is based on a vision of education that envisions organizing professional training workshops, technical support, and establishing teacher teams to allow for collaboration and general ICT use.

Like any other study, this one has its limitations. The findings are based on four mathematics lead teachers. It will be important to explore the experiences of all mathematics teachers in STEM, which may be characterized by different performances, and to develop measurable indicators for profiling STEM on a larger scale. In addition, the focus was on mathematics as a subject; it would be important to extend the study to other subjects like science, engineering, and technology.

## Conclusion

This chapter explored teachers' experiences of blended mathematics classrooms in MST academy hubs. Responses from qualitative data revealed that teachers learn from one another in a workshop and not only from the facilitator, especially in the pedagogical use of ICT resources. In particular, it emerged from data that the introduction and use of the online Dashboard, which is connected to the Department of Basic Education, assists teachers in retrieving information for the learners that can be used even when absent. Data also showed that STEM teachers were provided with laptops with pre-loaded learning material, and learners were supplied with tablets and data to access the pre-loaded MST material, making it easier for lead teachers to ensure improvement in STEM subjects. The challenges reported affected a few hubs where the manager in the school would take strict control of the e-learning tools and thus afford teachers no time to use them. This could be associated with vandalism, which was reported in one of the hubs where the smartboard was tampered with, and some laptops were stolen. However, these were only a few reported cases, and this challenge will be addressed when security measures are improved.

The disruptions caused by the COVID-19 pandemic to traditional teaching and learning in the physical classroom settings indicate the necessity to ensure that our schools respond to connectivity such that mathematics teachers can fully utilize both the technological, pedagogical and mathematics CK at all levels. Consequently, the STEM experiences of online connection to schools for the teaching and learning of mathematics made us realize that although the training of successful and effective mathematics teachers in online spaces during the pandemic was a complex and dynamic task, issues of social justice, quality, equity, and connectivity can assist in continuously improving mathematics teaching and performance, especially in a country as unequal as South Africa. Technology can influence STEM education in various ways; in particular, it can contribute to rethinking STEM pedagogy. The findings of this study can be replicated in other grades in the STEM hubs and other mainstream schools. The policy should enforce classrooms in which a blended learning method of instruction can be affected, especially in low-socioeconomic areas, so that teachers can assist learners with hands-on, independent, and self-motivated learning. This study was small-scale since it was conducted in one province on STEM subjects. It would be interesting to learn from large-scale studies the influence of ICT use on STEM subjects after the COVID-19 pandemic. Future studies can be conducted on how STEM teachers can be empowered to use blended instruction methods to ensure that their classrooms provide learners with first-hand experience of the content and take ownership of their learning.

## Bibliography

Angrist, N., Bergman, P., & Matsheng, M. (2021). *School's out: Experimental evidence on limiting learning loss using "Low-Tech" in a pandemic* (NBER Working Paper 28205). Updated January 2021. <http://www.nber.org/papers/w28205>.

Assareh, A., & Bidokht, M. H. (2011). Barriers to e-teaching and e-learning. *Procedia Computer Science 3*, 791–795.

Bingimlas, K. A. (2009). Barriers to the successful integration of ICT in teaching and learning environments: A review of the literature. *Eurasia Journal of Mathematics, Science, and Technology Education 5*(3), 235–245. <https://doi.org/10.12973/ejmste/75275>.

Birkinshaw, J., Brannen, M. Y., & Tung, R. L. (2011). From a distance and generalisable to up close and grounded: Reclaiming a place for qualitative methods in international business research. *Journal of International Business Studies 42*(5), 573–581. <https://doi.org/10.1057/jibs.2011.19>.

Chatterjee, S., Moody, G., Lowry, P. B., Chakraborty, S., & Hardin, A. (2020). Information technology and organisational innovation: Harmonious information technology affordance and courage-based actualisation. *The Journal of Strategic Information Systems 29*(1), 101596. <https://doi.org/10.1016/j.jsis.2020.101596>.

Chen, R., Liang, W., Jiang, M., Guan, W., Zhan, C., Wang, T., Tang, C., Sang, L., Liu, J., Ni, Z., Hu, Y., Liu, L., Shan, H., Lei, C., Peng, Y. Wei, L., Liu, Y., Hu, Y., Peng, P., …, & Medical Treatment Expert Group for COVID-19. (2020). Risk factors of fatal outcome in hospitalised subjects with coronavirus disease 2019 from a nationwide analysis in China. *Chest, 158*(1), 97–105.

Clark, A. (2010). *Transforming Children's Spaces: Children's and Adults' Participation in Designing Learning Environments.* Routledge.

Coetzee, J., Neneh, B., Stemmet, K., Lamprecht, J., Motsitsi, C., & Sereeco, W. (2021). South African universities in a time of increasing disruption. *South African Journal of Economic and Management Sciences 24*(1), 1–12. <http://dx.doi.org/10.4102/sajems.v24i1.3739>.

Creswell, J. W. (2014). *Qualitative, Quantitative, and Mixed Methods Approach*. Sage.

Cui, P., & Zheng, L. (2018) A meta-analysis of the peer evaluation effects on learning achievements in blended learning environment. In: Cheung, S., Kwok, L., Kubota, K., Lee, L.K., & Tokito, J. (Eds.) Blended learning: Enhancing learning success. pp. 227–237.

Ertmer, P. A. (1999). Addressing first- and second-order barriers to change: Strategies for technology integration. *Educational Technology Research and Development 47*(4), 47–61. <https://doi.org/10.1007/BF02299597>.

Gamede, B. T., Ajani, O. A., & Afolabi, O. S. (2022). Exploring the adoption and usage of learning management system as alternative for curriculum delivery in South African higher education institutions during COVID-19 lockdown. *International Journal of Higher Education*, *11*(1), 71–84. <https://doi.org/10.5430/ijhe.v1>.

Gumede, L., & Badriparsad, N. (2022). Online teaching and learning through the students' eyes—Uncertainty through the COVID-19 lockdown: A qualitative case study in Gauteng province, South Africa. *Radiography 28*(1), 193–198. <https://doi.org/10.1016/j.radi.2021.10.018>.

Hadijah, S., & Shalawati, S. (2017). Investigating teacher barrier to ICT (Information Communication Technology) integration in teaching English at senior high school in Pekanbaru. Proceedings of ISELT FBS Universitas Negeri Padang *5*, 302–310.

Helsa, Y., Darhim, D., Juandi, D., & Turmudi, T. (2021). Blended learning in teaching mathematics. *AKSIOMA: Jurnal Program Studi Pendidikan Matematika 10*(2), 733–743.

Indumathy, V., Shifamol, K., Beegammk, H., Laiju, L., & Aloysious, A. (2020). Pedestrian level of service at unsignalized intersections and junction improvement. *Journal of Dental and Medical Sciences 19*(7), 41–47.

Jethro, O. O., Grace, A. M., & Thomas, A. K. (2012). E-learning and its effects on teaching and learning in a global age. *International Journal of Academic Research in Business and Social Sciences 2*(1), 203.

Lawton, B., Brandon, P. R., Cicchinelli, L., & Kekahio, W. (2014). *Logic models: A tool for designing and monitoring program evaluations*. (REL 2014-007). Regional Educational Laboratory Pacific. <http://ies.ed.gov/ncee/edlabs/projects/project.asp?ProjectID=404>.

Lee, & J. Tokito (Eds.), *Blended Learning. Enhancing Learning Success*. ICBL 2018. Lecture Notes in Computer Science (Vol. 10949). Springer. <https://doi.org/10.1007/978-3-319-94505-7_18>.

Liang, S. W., Chen, R. N., Liu, L. L., Li, X. G., Chen, J. B., Tang, S. Y., & Zhao, J. B. (2020). The psychological impact of the COVID-19 epidemic on Guangdong college students: The difference between seeking and not seeking psychological help. *Frontiers in Psychology 11*, 2231. <https://doi.org/10.3389/fpsyg.2020.02231>.

Malterud, K., Siersma, V. D., & Guassora, A. D. (2016). Sample size in qualitative interview studies: Guided by information power. *Qualitative Health Research 26*(13), 1753–1760. <https://doi.org/10.1177/1049732315617444>.

McGarr, O., & McDonagh, A. (2021). Exploring the digital competence of pre-service teachers on entry onto an initial teacher education programme in Ireland. *Irish Educational Studies 40*(1), 115–128. <https://doi.org/10.1080/03323315.2020.1800501>.

Mishra, P., & Koehler, M. J. (2008). Technological pedagogical content knowledge: A framework for teacher knowledge. *Teachers' College Record 108*(6), 1017–1054. <https://doi.org/10.1111/j.1467-9620.2006.00684.x>.

Netshakhuma, N. S. (2020). Analysis of archives infrastructure in South Africa: Case of Mpumalanga provincial archives. *Global Knowledge, Memory and Communication*. <https://www.emerald.com/insight/2514-9342.htm>.

Nicholson, P. (2007). A history of e-learning. In B. Fernández-Manjón, J. M. Sánchez-Pérez, J. A. Gómez-Pulido, M. A. Vega-Rodríguez, J. Bravo-Rodríguez (Eds.). *Computers and Education* (pp. 1–11). Springer. <https://doi.org/10.1007/978-1-4020-4914-9_1>.

Patton, M. Q. (2015). *Qualitative Research and Evaluation Methods: Integrating Theory and Practice*. Sage.

Pule, K. G., & Ngoveni, M. A. (2024). Perceived effectiveness of online learning for mathematics pre-service teachers in a rural university during the Covid-19 pandemic. *International Journal of Social Science Research and Review 7*(2), 148–162. <https://doi.org/10.47814/ijssrr.v7i2.1853>.

Putra, A. K., Deffinika, I., & Islam, M. N. (2021). The effect of blended project-based learning with STEM approach to spatial thinking ability and geographic skill. *International Journal of Instruction 14*(3), 685–704. <https://doi.org/10.29333/iji.2021.14340a>.

Quadri, N. N., Muhammed, A., Sanober, S., Qureshi, M. R. N., & Shah, A. (2017). Barriers effecting successful implementation of e-learning in Saudi Arabian universities. *International Journal of Emerging Technologies in Learning (Online) 12*(6), 94. <https://doi.org/10.3991/ijet.v12i06.7003>.

Saldaña, J. (2014). Coding and analysis strategies. In P. Leavy (Ed.), *The Oxford Handbook of Qualitative Research* (pp. 581–605). Oxford University Press.

Schoepp, K. (2005). Barriers to technology integration in a technology-rich environment. *Learning and Teaching in Higher Education: Gulf Perspectives 2*(1), 56–79.

Shinas, V. H., Karchmer-Klein, R., Mouza, C., Yilmaz-Ozden, S., & Glutting, J. (2015). Analyzing preservice teachers' technological pedagogical content knowledge development in the context of a multidimensional teacher preparation program. *Journal of Digital Learning in Teacher Education 31*(2), 47–55. <https://doi.org/10.1080/21532974.2015.1011291>.

Shulman, L. S. (1986). Those who understand: Knowledge growth in teaching. *Educational Researcher 15*(2), 4–14.

Sointu, E., Valtonen, T., Cutucache, C., Kukkonen, J., Lambert, M. C., & Mäkitalo-Siegl, K. (2017). Differences in preservice teachers' readiness to use ICT in education and development of TPACK. In P. Resta, & S. Smith (Eds.), *Proceedings of Society for Information Technology & Teacher Education International Conference* (pp. 2462–2469). Association for the Advancement of Computing in Education (AACE). Retrieved June 6, 2024 from <https://www.learntechlib.org/p/177544>.

Sun, Z, Xie, K., & Anderman, L. H. (2018). The role of self-regulated learning in students' success in flipped undergraduate math courses. *Internet and Higher Education 36*, 41–53. <https://doi.org.10.1016/j.iheduc.2017.09.003>.

Tseng, M. L., Tan, R. R., & Siriban-Manalang, A. B. (2013). Sustainable consumption and production for Asia: Sustainability through green design and practice. *Journal of Cleaner Production 40*, 1–5. <https://doi.org/10.1016/j.jclepro.2012.07.015>.

UNESCO (2020). *How Many Students Are at Risk of Not Returning to School?* <https://unesdoc.unesco.org/ark:/48223/pf0000373992>.

Vincent-Lancrin, S., Cobo Romaní, C., & Reimers, F. (2022). *How Learning Continued During The COVID-19 Pandemic: Global Lessons From Initiatives To Support Learners And Teachers.* OCDE; The World Bank, 2022. <https://hdl.handle.net/11162/219986>.

Wahyudi, M., Mukrodi, M., Harras, H., & Sugiarti, E. (2020). Wirausaha Muda Mandiri: Learning, sharing & practice. *Scientific Journal of Reflection: Economic, Accounting, Management and Business 3*(1), 101–110. <https://doi.org/10.37481/sjr.v3i1.120>.

Zaharah, Z., Kirilova, G., & Windarti, A. (2020). Impact of coronavirus outbreak towards teaching and learning activities in Indonesia. *SALAM Jurnal Sosial dan Budaya Syar 7*(3) <https://doi.org/10.15408/sjsbs.v7i3.15104>.

Zydney, J. M., Warner, Z., & Angelone, L. (2020). Learning through experience: Using design-based research to redesign protocols for blended synchronous learning environments. *Computers and Education*, 143, 103678. <https://doi.org/10.1016/j.compedu.2019.103678>.

Zhao, X., Chen, J., Karczewicz, M., Zhang, L., Li, X., & Chien, W. J. (2016, March). *Enhanced multiple transform for video coding.* In 2016 Data Compression Conference (DCC) (pp. 73-82). IEEE.

CHAPTER 7

# Postgraduate Students' Experiences of Online Teaching of Science, Technology, Engineering, and Mathematics Subjects

Asheena Singh-Pillay[1]
[1]University of KwaZulu-Natal

**ABSTRACT**
Science, Technology. Engineering and Mathematics (STEM) subjects are a conduit for developing twenty-first-century skills. In developing countries, STEM subjects are often treated as isolated, stand-alone subjects, and teachers struggle to make connections across STEM subjects. Additionally, each STEM subject has its own nomenclature. In South Africa, the challenge is compounded by the language of instruction often differing from learners' home languages and the inherent complexity of STEM concepts. This chapter argues that information and communication technologies (ICT) can address the language challenges learners and teachers encounter with STEM subjects and promote disciplinary science understanding by linking crosscutting concepts. Mishra and Koehler's Technological Pedagogical and Content Knowledge model famed this interpretative case study research conducted at Richmond University in KwaZulu-Natal, South Africa. This research answered the following questions: What are postgraduate STEM students' experiences of online teaching and learning during the COVID-19 pandemic, and its impact on their teaching practice? Data were generated via discussion forums and reflective journals from 20 purposively selected postgraduate STEM teachers enrolled for the honors module Research in Technology, which required them to be *au fait* in using technology. All ethical protocols were adhered to. The findings highlight the challenges STEM teachers encountered during online teaching and the opportunities they experienced when using ICT for translanguaging and supporting students in developing disciplinary science understanding by linking crosscutting concepts. The novel findings of this study illustrate how to enact STEM teaching in developing countries, address issues of equity, and leverage ICT to enhance practice in STEM subjects post-COVID.

*Keywords:* Crosscutting connections, ICT, online teaching, STEM subjects, TPACK, translanguaging, twenty-first century skills

## Introduction

The COVID-19 pandemic disrupted our everyday routine. Sanitizing, wearing masks, social distancing and remote online teaching and learning became the new normal. With the transition to online teaching, all teachers, irrespective of their competence in using technologies, were required to integrate technologies in teaching, designing and developing online lessons and assessments (Farmer and West, 2019). However, students and teachers require training to use technology for teaching and learning (Naidoo and Singh-Pillay, 2021). Within the South African context, the move to online teaching and learning was not a seamless process. It was assumed that the move to online teaching entailed favorable living conditions, food security for families, conducive learning spaces, devices to engage in online teaching and learning and stable uncapped internet connectivity (Singh-Pillay and Naidoo, 2020). Over and above these underlying assumptions, science, technology, engineering, and mathematics (STEM) teachers were still expected to engage learners and develop key skills such as critical thinking, entrepreneurship, communication, collaboration, decision-making, leadership, problem-solving, responsibility, creativity and innovation while teaching STEM subjects. The first implication is that STEM teachers are au fait with the content knowledge (CK) and online pedagogy associated with STEM subjects to develop the above mentioned skills. Second, STEM teachers should be able to integrate these four disciplines rather than teaching these disciplines in isolation. DeCoito and Richardson (2018) and Faloon (2020) document that using mobile digital technologies and information and communication technologies (ICT) impacts STEM teachers' competencies to promote conceptual and skills development in their learners.

It is crucial to explore postgraduate STEM students' experiences of online teaching and learning during the COVID-19 pandemic and its impact on their teaching practice, as online teaching was a novel experience for many of them. This chapter comprises six sections: introduction, literature review, theoretical framework, methodology, findings, discussion, and conclusion.

## Literature Review

This section interrogates three aspects: the use of technology in STEM teaching and learning, the use of technology to overcome language barriers to STEM teaching, and STEM teachers' online teaching experience.

### Utilization of technologies in STEM

DeCoito and Richardson (2018) emphasized the value of using digital technologies for STEM teaching and learning. These scholars elucidate how students in STEM classrooms can benefit from learning via digital technologies through practical work, simulations, modelling, and data analytics. The benefits include improvement in students' visual processes, the connection of crosscutting concepts, problem-solving, application of learning to real-life contexts, collaboration, organizing data for graphing, and manipulating variables, provided teachers are trained to use these technologies in their teaching.

Digital technologies are not just for teaching but also for online assessments. Barril (2018) avers that if teachers embark on online teaching, they must be trained to create multiple innovative online assessments. Faber et al. (2017) emphasize that prompt corrective feedback is essential for learners to identify errors, obtain better cognitive understanding and enhance performance. Despite these benefits, the challenges associated with using technologies in STEM cannot be ignored. For instance, the digital divide, the digital use divide, and the infrastructure among developed and developing nations (Cheshmehzang et al., 2022) complicate incorporating technologies in school education in developing countries (Lubega et al., 2021). Additionally, teachers' self-efficacy and training impact the integration of technologies into teaching STEM subjects.

## Technologies to diminish language dynamics in STEM

The language of the colonizer is still dominant in post-colonial global South contexts and is embedded in the national language policies as the official medium of instruction (Mthombeni and Ogunnubi, 2021; Nhongo and Tshotsho, 2021). In many global South countries, STEM teachers have the added duty to facilitate STEM terminology and STEM content in the medium of instruction (English) and cope with the learners' home language (Semoen and Mutekwe, 2021). STEM teachers have the extra responsibility of being teachers of language whilst being expert pedagogues in STEM language use.

Learners entering STEM classrooms carry their sociocultural background with them, which impacts their learning. Mokiwa (2020) assert that the disconnect between the medium of instruction and the learners' home language accounts for these learners' underachieving in STEM subjects. Students and teachers whose home language is not the same as the medium of instruction have to deal with teaching and learning support resources written from a Eurocentric stance, which is a stumbling block to accessing information. STEM teachers do not receive professional development support to provide them with skills and knowledge to aid and assist the indigenous learners' language needs in the STEM classroom.

Learners whose home language differs from the medium of instruction experience challenges or barriers during their encounter with STEM subjects. This discrepancy makes one wonder how the multiple languages of the local community are catered to by using ICT to facilitate the meaning-making of STEM text, acquiring content and process skills. In contrast, learners use their home language to access information. Technology can significantly enhance educational experiences and overcome various barriers in diverse classroom settings regarding the digital divide, use, and accessing CK. For example, Translation and Language Learning Apps like Google Translate and Duolingo help students who speak different languages understand classroom material and improve their language skills. Subtitling and Transcription Services, such as YouTube and Microsoft Stream, provide automatic subtitles and transcriptions, making video content more accessible to non-native speakers. Interactive and Gamified Learning Apps like Kahoot! and Quizlet make learning interactive and fun, catering to students who benefit from engaging game-like environments.

The value of translanguaging pedagogy as leverage to minimize language barriers prevalent in the STEM classroom is highlighted by Karlsson, Larsson, and Jakobsson (2019) and Nhongo and Tshotsho (2019). This means that the home language is an invaluable asset that ought to be used for teaching and assessments. Feedback from indigenous learners ought to inform teachers' instructional planning and design. To espouse translanguaging pedagogies and embrace their learners' linguistic backgrounds, STEM teachers must be capacitated to use translanguaging digital and ICT resources. The type and kind of training teachers need to use translanguaging digital resources cannot be a one-size-fits-all training model; the training will depend on the context and technological proficiency (Ho and Tai, 2021).

### STEM teachers' experiences on online teaching under COVID conditions

The rapid shift to online teaching during the COVID-19 pandemic resulted in many challenges related to variability in teachers' subject matter, pedagogy, technological proficiency, student factors and institutional support, and how teachers responded to students and assessments during the transition to online teaching. Barron Rodriguez et al. (2021), argue that these modifications increased the pressure on already burdened teachers expected to interface online content material with their existing module templates and syllabi while providing pastoral care to their students and navigating a work-life balance. According to DeCoito, 2020, teachers could not transfer face-to-face materials to an online learning environment in some subjects without making significant revisions. For example, teachers found it hard to nurture higher-order thinking, implement student-centered teaching methods, and engage students in hands-on inquiry-based activities and laboratory experiments. Doucet et al. (2020), reported that teachers complained about the poor training received for online teaching, lack of training in online pedagogies, balancing home life with work life, coping with isolation, inability to design activities that promote deep learning and sustain learners and poor network connectivity.

## Theoretical Framework

Mishra and Koehler's (2006) Technological Pedagogical Content Knowledge (TPACK) is a theoretical framework that explains integrating technology (TK), pedagogy (PK), and CK needed for successful teaching. TPACK is the knowledge teachers ought to have to incorporate technology into pedagogy and CK for effective teaching and learning. Shulman's (1986) notion of pedagogical content knowledge (PCK) undergirds the TPACK model.

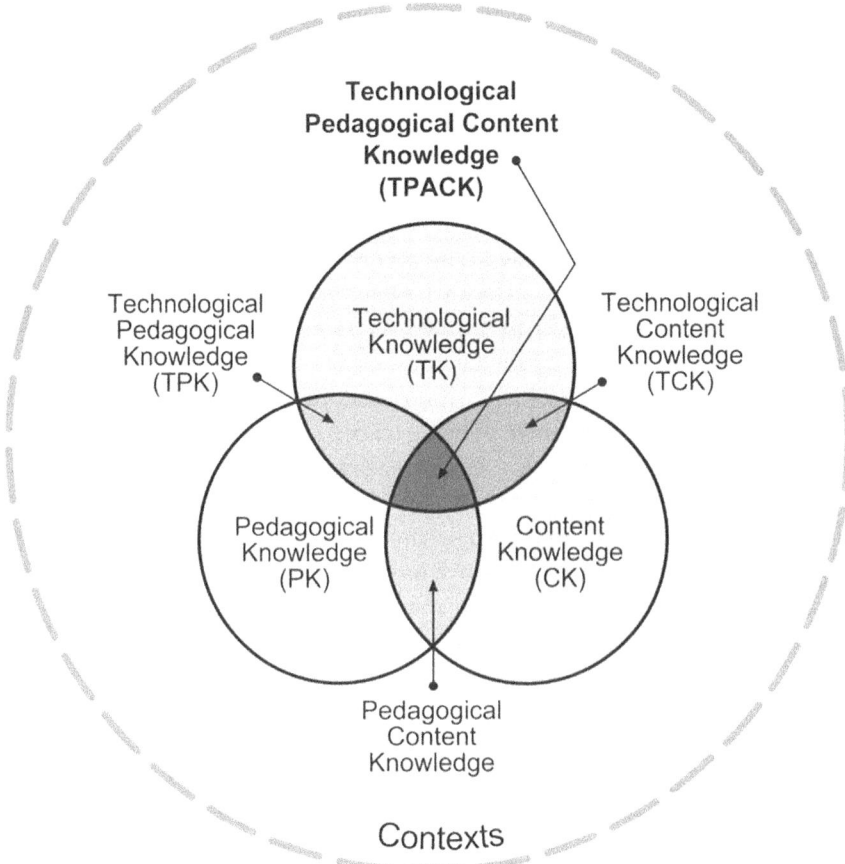

Figure 7.1: The components of the TPACK framework.

Technology knowledge (TK) is information about technologies and the skills needed to use certain technologies (Koehler et al., 2014). Due to constant technological upgrades, a person's TK needs to evolve with such advancements. Technological content knowledge (TCK) is how to link technology and content to change the subject matter. Technological pedagogical knowledge (TPK) is the know-how or expertise of the fitness of different technologies used for teaching and learning. It also involves cognizing how teaching pedagogies may be modified using specific technologies (Greenhow et al., 2021). TPACK amalgamates content, pedagogy, and technology knowledge. TPACK depends on levels of pre-existing teacher knowledge and the technological support teachers require to reach this level of integration. In a way, TPACK considers the complexities and practical challenges teachers might face in integrating technology, pedagogy, and CK in natural classroom settings. TPACK arises from teaching with technology using appropriate pedagogical techniques in student-centered settings (Bibi and Khan, 2017).

## Methodology

This qualitative interpretative case study research was conducted in semester 2 of 2020 when online teaching and learning occurred during pandemic conditions. Twenty STEM students registered for their Honors degree (research in technology module) at Richmond University formed the sample for this study. The hons degree is considered as a postgraduate study. Hence, the participants are referred to as postgraduate STEM students. Purposive sampling was used in this study, and participants met the requirement of being teachers of STEM subjects at schools. Protocols concerning ethics that were in keeping with Richmond University requirements were adhered to. Due to lockdown and social distancing conditions, data was generated electronically via online discussion forums (3) and an electronic reflective journal. Interactive online discussion forums are beneficial for generating data, as they allow for shared dialogue and for responses to be probed. The questions posted on the discussion forum are reflected in Table 7.1.

Table 7.1: Discussion forum questions

| Question | Sub questions |
|---|---|
| What were the participants' experiences of online teaching and learning during COVID-19? | - What are your experiences of online teaching and learning under these conditions?<br>- Did you encounter any challenges with online teaching and learning?<br>- What steps can be taken to improve your experience of online teaching and learning?<br>- Are there any benefits associated with online teaching and learning under COVID conditions that can be used post-COVID to teach STEM subjects? |
| How do these experiences impact their teaching practice? | - What changes, if any, occurred in your TK, TCK, TPK, and TPACK?<br>- Has your teaching been shaped by your experiences of online teaching and learning? |

Participants were advised to maintain an electronic reflective journal by noting the opportunities and challenges they encountered during online teaching. This study had a one hundred per cent participation rate. Before data could be collected, participants selected a pseudonym, P1–20, to identify their participation during data generation. The discussion forum data were transcribed before analysis could begin.

Before coding could begin, the transcripts and reflective journals were read many times to group similar codes into themes. Content analysis was used, and the constructs from the TPACK framework were used where possible to make sense of the data. The codes that emerged from the data and their respective themes are reflected in Table 7.2.

Table 7.2: Codes and themes

| Codes | Sub-themes | Themes |
|---|---|---|
| Over hall, change, readjust, calibrate | Recalibrate teaching | Challenges with online teaching and learning |
| Cross-language borders, code switch, improved performance | Translanguaging | Opportunities with online teaching |
| Virtual practical, virtual field trips, simulation of process skills | Simulations | |
| Make connections across STEM, integrate similar concepts in STEM, link concepts across STEM | Link crosscutting concepts | |

Two themes emerged regarding postgraduate STEM students' experiences of online teaching: the challenges of online teaching and learning and the opportunities that impacted their teaching.

## Findings and Discussion

Data from the discussion forum and reflective journal were used to answer the research question: What are postgraduate STEM students' experiences of online teaching and learning during the COVID-19 pandemic and its impact on their teaching practice? Data analysis reveals two themes: challenges with online teaching and opportunities with online teaching. Each theme is presented and discussed next.

### Challenges with online teaching and learning
#### Recalibration of teaching
The data from the first discussion forum and initial entries in the reflective journals indicate that all participants encountered challenges with online teaching and

learning regarding their use of technologies, knowledge of which technologies to use, pedagogy, and assessments, as is visible in the excerpts below.

> *I am frustrated. My job as a teacher has changed because of the pandemic lockdown conditions. I am expected to teach and assess online, but I lack confidence in using technology to teach. I was not trained for online teaching, and I am failing to use the right strategies for engaging in online lessons and assessments in science and technology classrooms.* P7, Discussion forum

> *I feel overwhelmed with online teaching because face-to-face teaching is so much easier. I do not know how to research my students cognitively on the online platform. All aspects of my teaching, pedagogy, assessment, and content must be readjusted.* P1, Discussion forum

> *I am uncertain if I am using the appropriate technologies to teach the different physics and Life Sciences sections and if learners benefit from online learning. I second guess the teaching strategy I am using- you cannot just transfer face-to-face teaching methods and assessments to the online platform. Practical work is a considerable challenge.* P12, Reflective journal

The abrupt move during the pandemic to online teaching forced teachers to rethink teaching and learning via the use of technologies. Initially, the lack of appropriate in-depth training for online teaching created uncertainties among these 20 postgraduate STEM participants. The uncertainties participants experienced impacted their confidence in teaching with technologies, their skills in using technology, designing appropriate lessons and assessments, reaching learners cognitively, and conducting practical work in STEM subjects. In short, participants believed there is a dire need to readjust and recalibrate all aspects of their teaching about pedagogy and assessments due to their lack of confidence in teaching with and through technology. The above findings concur with the study conducted by Baptista et al. (2020), which indicates that when the pandemic first broke out, STEM teachers did not know how to implement remote online instruction. Reimers (2022) studied teaching and learning in 14 countries and highlighted that teachers found it challenging to effectively use technologies and develop online lessons and assessments.

Training in the use of technology is a key factor that impacts teachers' ability to rethink teaching with technology and their confidence to use technology. The participants in this research initially explicitly bemoan their lack of knowledge on which technology to use (TK) and how to use the technology to create new representations for specific content (TCK). This means that the absence of technological knowledge (TK) impacts their appropriate and effective use of technology to promote learning (TCK). Furthermore, they relied on the pedagogies and assessments they used during face-to-face teaching. This means participants initially struggled to embrace new and relevant pedagogies to teach with technology (TPK), which led them to realize the need to readjust their teaching. The above finding is in sync with Roussinos and Jimoyiannis's (2019) study, which emphasized that teachers could not effectively incorporate their PK into online teaching-learning due to the disconnect between their digital technology skills and knowledge of effective teaching approaches. The lack of enthusiasm initially to embark on online teaching has compromised participants' ability to integrate technology, content, and pedagogy and impacted their TPACK and skills needed to create meaningful, stimulating online lessons. Therefore, it can be argued that teachers need to be capacitated on using digital technologies through authentic, inquiry, and collaborative learning activities to create advanced learning environments that support the process of actively building students' knowledge (Roussinos and Jimoyiannis, 2019).

## Opportunities with online teaching

Analysis of data from the second and third discussion forums and subsequent entries in the reflective journal reveal that STEM postgraduate students were able to use online platforms quickly and encountered three types of opportunities: translanguaging, simulations, and integration of STEM content. Each subtheme is discussed next.

## Translanguaging

The terminology used in STEM subjects is abstract, foreign and very different from learners' everyday spoken home language. The contestation between STEM terminology, the learners' home language and the language of instruction hampers learners' ability to acquire information, participate, understand and perform well. The excerpts below reveal the challenges learners encounter in STEM classrooms:

> *The terms and language used in Chemistry, Life Sciences and Physics are difficult, foreign and abstract. Learners from indigenous backgrounds who do not speak English at home cannot make sense of these tricky words. It is not that they are stupid or backwards. The abstractness of the terms prevents them from accessing learning. It is all about overcoming a language barrier. The workshops on technologies for translanguaging, such as Google Classroom, voice typing, and Google Dictionary, have altered my outlook and teaching strategies. I use these technologies to help learners learn STEM through their home language.*
> P6, Discussion forum

> *I am comfortable crossing language barriers in my classroom, and I have embraced translanguaging to support my learners and amended the instructional design of my lessons, assessments, and*

> *rubrics. The outcome is excellent. Indigenous learners are keen to learn and actively participate in the lesson and learning. There has been a marked improvement in learners' achievement and test scores.* P14, Reflective journal

> *My teaching has improved after I learned how to use technology for translanguaging. I am confident in using technology when teaching difficult, abstract content in chemistry and physics. Learners can switch codes between their home language, STEM language, and English to write reports on practical work using voice typing features and Google Dictionary. Changing my teaching approach has made a difference to my learners and their engagement in lessons, and the bonus is that their performance has improved. I wish I had learned about translanguaging earlier.* P5, Reflective journal

The testimonies above highlight participants' positive experiences of the training received to use translanguaging technologies (Google Classroom, voice typing, and Google Dictionary) in their teaching of STEM subjects. The training changed their TK, TPK, TCK and TPACK and how they teach STEM effectively through the use of technologies. They use translanguaging pedagogies effortlessly to address the age-old language barriers Indigenous learners encounter in STEM classrooms. These technologies have made it possible to use the learners' linguistic backgrounds as a significant part of classroom practice. The benefits of translanguaging technology, such as ease of switching between the home, medium of instruction, and STEM language, are visible in the above findings.

Further, the findings illustrate how technology can be used to embrace a humanizing approach and facilitate inclusive experiences for indigenous learners, who can comfortably turn to their home language to make sense of STEM concepts and content. In such instances, the home language is not considered backwards or inferior to the LOLT nor hinders accessing STEM knowledge/learning. The language contestation, dilemmas and power struggles in STEM classrooms (between the LOLT and learners' home language), as well as issues of access and social justice, can be addressed via translanguaging.

The findings of this study align with those of Zano and Charamba's (2021) study. They observed improved learners' participation and achievements when translanguaging pedagogies were used. Moreover, the results of this study clarify that the learners' home language can be used as a platform to promote understanding of STEM subjects and scaffold learning if translanguaging approaches are used. This finding is worth noting, as it is antithetical to Dlamini's (2020) study, which emphasized that learners whose home language is not English struggle with making sense of STEM knowledge and consequently perform poorly in STEM subjects. Dlamini backs her claim by referring to learners' performance in the TIMMS study. The findings of this study clarify that translanguaging can be an asset for teaching and learning in post-COVID STEM classrooms. Poza (2019) elucidates the benefits of translanguaging in STEM classrooms. These benefits include support of language development, use of scientific discourse, higher-order thinking, conceptual understanding, and creating a more humane and inclusive learning environment for learners whose home language is not English.

## Simulations

Practical work, demonstration, hypothesis testing, and experiments are the core of STEM teaching and learning. They allow for clarification of concepts, linking theory with practice, and developing process skills (Flores and Marzullo, 2021). Access to laboratories was not possible during the lockdown conditions of the COVID-19 pandemic, so alternative means had to be found to involve students in practical work. The testimonies below highlight the alternative ways students were involved in practical work.

> *I discovered that simulations are interactive and promote the development of process skills, critical thinking, creativity and inquiry-based learning. It is the best way to conduct practical work without a functional laboratory. PhET simulations were used for experiments, explaining abstract concepts and involving student activity in manipulating apparatus, variables, and collaborations. Simulations work well in physics, math and EGD. P1, Discussion forum*

> *My life Sciences learners went on field trips, virtually, to aquatic ecosystems (dams, rivers, lakes and sea). Learners explored these ecosystems, made observations, gathered information on biotic and abiotic factors, and created food chains, webs, and ecological pyramids. After their virtual field trip, learners were required to present their findings via Zoom. The advantage of using simulations was that the field trip was free for the learners, and there were no safety and security issues. Learners reported that the field trip was fun, free and an exciting hands-on experience. P11, Discussion forum*

> *Initially, I conducted experiments, recorded them and shared them on the learn portal. The learners were not actively involved. They just observed. As my confidence grew, I used YouTube videos, and finally, I got learners involved in kitchen chemistry on Zoom. They used household items to conduct practicals at home, such as neutralisation, pH, and indicators. Learners were actively involved in the testing, making, conducting experiments and developing process skills. Breakaway rooms were created for student discussions and the presentation of findings. P19, Reflective journal*

The strategies and approaches deployed to involve learners in practical work when laboratories were inaccessible during the pandemic are visible in the above excerpts. Learners could be involved with investigating, predicting, evaluating, developing explanations, communicating, thinking critically, problem-solving, collaborating, being creative, and innovating rather than just recording and observing the experiment. The above findings align with Aliyu and Talib's (2019) study, which shows that in the absence of a laboratory or chemicals and functional lab equipment (Rani et al., 2018), learners can still engage in practical work. These findings exemplify that a virtual platform can support experiential learning about practical work and abstract concepts, and acquire process and twenty-first-century skills. Further, it can be inferred that if teachers are trained and knowledgeable about the relevant technologies and the accompanying pedagogies, their TPACK will increase, and their practice will change (Lestari and Supahar, 2020). The above findings are relevant for conducting practical work in a post-COVID context in schools that lack physical, human, and financial resources. Ramnarain explains that in South Africa, practical work is lacking in STEM subjects, especially in rural settings. He sees the absence of practical work as a lost opportunity to learn science content and develop process skills. Conducting practical work virtually means that practical work can be conducted anywhere, at any time, in a safe, accident-free zone.

### Linking crosscutting concepts

STEM subjects are regarded as standalone and are taught as separate subjects. This means that common crosscutting concepts are not linked to advanced conceptual understanding. With the innovative use of technology, Postgraduate STEM students learnt to make connections amongst crosscutting concepts in STEM disciplines, as is evident in the excerpts below:

> All thanks to technology, I now craft my lessons differently. I draw on the design process, inquiry-based, P21 framework for 21st-century skills when teaching vectors in physics. Students are expected to link learning on dimensions from mathematics and spatial visual manipulation from EGD. Thus, I connect cross-cutting concepts and learning in STEM subjects. The discussion forum has a space to share ideas and tried and tested practices. It is a non-judgmental, safe space to learn. P11, Discussion form

> When I teach acid and bases in chemistry, I now forge ties to Life Sciences, the concept of neutralisation, and the use of antacids during indigestion. Also, connections are forged with mathematics, particularly logarithms, to calculate pH values. Using technologies has empowered me to link crosscutting concepts in STEM. P4, Discussion forum

> When I teach, I no longer think of math, physics, life science, and EGD as separate subjects whose content cannot be linked to scaffolding and supporting conceptual learning. Using technology, I draw on the broad base of STEM subjects to facilitate teaching and learning. I have broken free from the shackles of compartmentalised teaching and use technology to see connections between crosscutting concepts and apply them in my teaching. When teaching about diffraction and X-rays in physics, I link it to radioactivity

in chemistry and contextual issues in the community, like why pregnant ladies need to wear a lead apron when taking an X-ray and the need to disclose their pregnancy status when going for X-rays and scans. Online teaching is good; I have learnt so much about technologies that can be used to enhance teaching and learning. I am confident about my online pedagogy, design of lessons and material development. P15, Reflective journal

The shift in participants' confidence to teach with technology is visible in the above testimonies compared to their testimonies in Section 5.1.1. Their improved TCK, TPK, and TPACK influenced their teaching practice. Their teaching approach, support and guidance provided to students evolved from a silo mentality of STEM subjects (that is, treating subjects as discreet isolated learning areas) to teaching STEM subjects in an integrated way, where links are formed between crosscutting concepts across the STEM subjects. Improvement in knowledge and confidence in using technologies is quintessential for teachers to make a paradigm shift in pedagogy and linking of crosscutting concepts in STEM subjects.

The finding brings to the fore the value of having a safe space (via the discussion forum) during a learning curve, where uncertainties and tried and tested approaches to teaching can be shared. This finding is in sync with that of Gehrke and Kezar (2017) study which reported on the value of a community of practice where teachers are free to share and exchange ideas about their teaching, what works, what does not work, why it does not work so they can collectively enhance their learning on digital technologies, online pedagogy, teaching STEM differently by integrating STEM concepts.

## Conclusion

Postgraduate STEM students' experiences of online teaching and learning during the COVID-19 pandemic and their impact on their teaching practice were explored in this study. The finding reveals that postgraduate STEM students initially experienced challenges with online teaching and learning. Similarly, Baptista et al. (2020) reported that STEM teachers did not know how to implement remote online instruction when the pandemic first broke out. The challenges that postgraduate STEM students encountered were linked to their lack of knowledge of technology (TK), knowledge of which technologies to use when (TC), and how (TPK), lack of knowledge on instructional design and pedagogy, and assessments. Participants encountered positive experiences with online teaching regarding choices of resources and technologies, and planning and designing tasks to facilitate active learning in STEM disciplines once their confidence and forms of knowledge (TPACK, TK, TC, and TPK) in the use of technology improved.

Participants' experiences related to the opportunities with online teaching impacted their practice. To resolve the divide between STEM language, the medium of instruction and the learners' home language, participants embraced translanguaging pedagogies in order to use the learners' home language as an invaluable resource to facilitate learning and catapult an inclusive, respectful, and socially just learning environment. The findings of this study concretize the space and place of translanguaging pedagogies and technologies in STEM classrooms in a post-COVID era, in post-colonial countries, where the learners' home language is estranged from the medium of instruction. Embracing and espousing a translanguaging pedagogy in the post-COVID era recognizes the cultural and language strength Indigenous learners bring into the STEM classroom. Furthermore, translanguaging can be used to address the issues of language power dynamics, equity, access, and social justice that have plagued STEM classrooms since the dawn of colonization. Proponents of translanguaging, Charamba (2023) and Vogel and García (2017), make a humble appeal to STEM teachers to allow Indigenous learners to rely on their home language for meaning-making in the classroom.

The various ways learners were engaged in practical work on a virtual platform reveal that in the absence of laboratories and reagents, innovative active learning can occur where learners can develop complex processes and twenty-first-century skills. This finding dispels the age-old cry that fitted laboratories are needed to be able to conduct practical work. The findings on practical work are noteworthy. They demarcate a path that could be taken to deal with the perpetual obstacles facing many rural and under-resourced South African schools in terms of limited financial and infrastructure resources for laboratories. Ramnarain (2020) and Bantwani (2017) assert that in rural schools, the teaching and learning of science subjects are affected by the lack of fully fitted and functional laboratories. Some values of virtual practical work are safety, the absence of faulty lab apparatus, and greater equity of access to quality resources.

Participants' ability to link crosscutting STEM concepts in their teaching was intrinsically connected to their improved TPACK and their potential to not envisage STEM subjects as having fixed boundaries. Kelly et al. (2021) noted that teachers who encountered positive experiences in the use of technology were able to link the application of knowledge to real contextual issues.

The findings of this study are important in a post-COVID era for the teaching of STEM subjects in Africa and the global South, the training of pre-service teachers, and continuous teacher professional development for in-service teachers. The findings advocate for using technologies for translanguaging, practical work, and linking crosscutting concepts in STEM teaching and learning.

## Recommendation

Further research is needed on how students' learning can be supported in STEM subjects by translanguaging technology at other institutions of higher learning. Studies could be conducted in rural settings on using PhET simulations to conduct practical work, promote understanding of difficult concepts and develop process and twenty-first-century skills.

## Bibliography

Aliyu, F., & Talib, C. A. (2019). Virtual chemistry laboratory: A panacea to problems of conducting chemistry practical at science secondary schools in Nigeria. *International Journal of Engineering and Advanced Technology (IJEAT)* 8(5C), 544–549. <http://dx.doi.org/10.35940/ijeat.E1079.0585C19>.

Bantwini, B. (2017). Analysis of teaching and learning of natural sciences and technology in selected eastern cape province primary schools, *South Africa Journal of Education* (67), 39–64. <https://doi.org/10.17159/2520-9868/i67a02>.

Baptista, J., Stein, M. K., Klein, S., Watson-Manheim, M. B., & Lee, J. (2020). Digital work and organisational transformation: emergent digital/human work configurations in modern organisations. *Journal of Strategic Information Systems* 29(2), 101618–101622. <https://doi.org/10.1016%2Fj.jsis.2020.101618>.

Barril, L. (2018). Assessment for culturally inclusive collaborative inquiry-based learning. In *Handbook of Distance Education* (pp. 311–320). Routledge.

Barron Rodriguez, M. R., Cobo Romani, J. C., Munoz-Najar, A., & Sanchez Ciarrusta, I. A. (2021). *Remote Learning During the Global School Lockdown: Multi-country Lessons* (English). Washington, DC: World Bank Group. <https://documents.worldbank.org/curated/en/668741627975171644/Remote-Learning-During-the-Global-School-Lockdown-Multi-Country-Lessons/>. Accessed October 7, 2023

Bibi, S., & Khan, S. H. (2017). TPACK in action: A study of a teacher educator's thoughts when planning to use ICT. *Australasian Journal of Educational Technology* 33(4), 70–87. <https://doi.org/10.14742/ajet.3071>.

Charamba, E. (2023). Translanguaging as bona fide practice in a multilingual South African science classroom. *International Review of Education*. <https://doi.org/10.1007/s11159-023-09990-0>.

DeCoito, I. (2020). The case for digital timelines in teaching and teacher education. *International Journal of E-Learning & Distance Education* 35(1), 1–36.

DeCoito, I., & Richardson, T. (2018). Teachers and technology: Present practice and future directions. *Contemporary Issues in Technology and Teacher*

*Education 18*(2), 362–378. <https://www.learntechlib.org/primary/p/180395/ accessed 7 October 2023>.

Dlamini, S. M. (2020). *Language policy and practice at a secondary school in Manzini: The case of teaching and learning* [Unpublished thesis]. UKZN.

Doucet, A., Netolicky, D., Timmers, K., & Tuscano, F. J. (2020). Thinking about pedagogy in an unfolding pandemic (An Independent Report on Approaches to Distance Learning during COVID-19 School Closure). *Work of Education International and UNESCO.* <https://issuu.com/educationinternational/docs/2020_research_covid-19_eng>.

Faber, J. M., Luyten, H., & Visscher, A. J. (2017). The effects of a digital formative assessment tool on mathematics achievement and student motivation: Results of a randomized experiment. *Computers & Education 106*, 83–96. <https://psycnet.apa.org/doi/10.1016/j.compedu.2016.12.001>.

Falloon, G. (2020). From digital literacy to digital competence: The teacher digital competency (TDC) framework. *Education Technology Research Development 68*, 2449–2472. <https://doi.org/10.1007/s11423-020-09767-4>.

Farmer, T., & West, R. (2019). Exploring the concerns of online K-12 teachers. *Journal of Online Learning Research 5*(1), 97–118. <https://www.learntechlib.org/primary/p/184482/>.

Flores, D. P., & Marzullo, T. C. (2021). The construction of high-magnification homemade lenses for a simple microscope: An easy "DIY" tool for biological and interdisciplinary education. *Advances in Physiology Education 45*(1), 134–144. <https://doi.org/10.1152/advan.00127.2020>.

Fuad, M., Ariyani, F., Suyanto, E., & Shidiq, A. S. (2020). Exploring teachers' TPCK: Are Indonesian language teachers ready for online learning during the COVID-19 outbreak? *Universal Journal of Educational Research 8*(11B), 6091–6102. <https://doi.org/10.13189/ujer.2020.082245>.

García, G., Johnson, S. I., & Seltzer, K. (2017). *The translanguaging classroom: Leveraging student bilingualism for learning.* Philadelphia, PA: Caslon.

Garcia, O. (2014). Translanguaging as normal bilingual discourse. In Hesson, S., Seltzer, K., & Woodley, H.H. *Translanguaging in curriculum and instruction: A CUNY-NYSIEB guide for educators.* CUNY Graduate Center.

Gehrke, S., & Kezar, A. (2017). The roles of STEM faculty communities of practice in institutional and departmental reform in higher education. *American Educational Research Journal 54*(5), 803–833. <https://doi.org/10.3102/000283121770673>.

Greenhow, C., Lewin, C., & Willet, K. B. (2021). The educational response to COVID-19 across two countries: A critical examination of initial digital pedagogy adoption. *Technology, Pedagogy and Education 30*, 7–25. <http://dx.doi.org/10.1080/1475939X.2020.1866654>.

Karlsson, A., Nygård Larsson, P., & Jakobsson, A. (2020). The continuity of learning in a translanguaging science classroom. *Cultural Studies of Science Education 15*, 1–25. <https://doi.org/10.1007/s11422-019-09933-y>.

Kelly, P., Hofbauer, S., & Gross, B. (2021). Renegotiating the public good: Responding to the first wave of COVID-19 in England, Germany and Italy. *European Educational Research Journal 20*(5), 584–609. <https://doi.org/10.1177/14749041211030065>.

Koehler, M. J., Mishra, P., Kereluik, K., Shin, T. S., & Graham, C. R. (2014). The technological pedagogical content knowledge framework. In *Handbook of Research on Educational Communications and Technology* (pp. 101–111). Springer, New York, NY.

Lestari, D. P., & Supahar, S. (2020). Students and teachers' necessity toward virtual laboratory as an instructional media of 21st-century science learning. *Journal of Physics: Conference Series 1440*, 012091. <https://doi.org/10.1088/1742-6596/1440/1/012091>.

Mishra, P., & Koehler, M. J. (2006). Technological pedagogical content knowledge: A framework for teacher knowledge. *Teachers College Record 108*(6), 1017–1054. <https://psycnet.apa.org/doi/10.1111/j.1467-9620.2006.00684.x>.

Mokiwa, H. (2020). The pedagogy of learning and teaching science in a multilingual classroom: Teachers' perspectives. *Africa Education Review 17*(4), 87–103. <http://dx.doi.org/10.1080/18146627.2020.1868075>.

Mthombeni, Z., & Ogunnubi, O. (2021). A socio-constructivist analysis of the bilingual language policy in South African higher education: Perspectives from the University of KwaZulu-Natal. *Cogent Education 8*(1), 1954465. <https://doi.org/10.1080/2331186X.2021.1954465>.

Naidoo, J., & Singh-Pillay, A. (2021). Online teaching and learning within the context of COVID-19: Exploring the perceptions of postgraduate mathematics education students. *Mathematics Education Journals 5*(2), 102–114. http://dx.doi.org/10.22219/mej.v5i2.17015>.

Nhongo, R., & Tshotsho, B. P. (2020). The problematics of language-in-education policies in post-independence in Zimbabwe. *Journal of Asian and African Studies 56*(6), 1304–1317. <https://doi.org/10.1177/0021909620962529>.

Poza, L. (2019). Where the true power resides: Student translanguaging and supportive teacher dispositions. *Bilingual Research Journal 42*(1), 1–24. <https://doi.org/10.1080/15235882.2019.1682717>.

Ramnarain, U. (2020). Inquiry-based learning in South African schools. In *School Science Practical Work in Africa* (pp. 1–13). Routledge. <http://dx.doi.org/10.4324/9780429260650-1>.

Rani, S. A., Mundilarto, W., & Dwandaru, W. S. B. (2019). Physics virtual laboratory: An innovative media in 21st-century learning. *Journal of Physics: Conference Series 1321*, 022026. <https://doi.org/10.1088/1742-6596/1321/2/022026>.

Reimers, F. (2022). Learning from a pandemic: The impact of COVID-19 on education around the world. In F. M. Reimers (Ed.), *Primary and Secondary Education During COVID-19: Disruptions to Educational Opportunity During a Pandemic* (pp. 1–37). <https://doi.org/10.1007/978-3-030-81500-4_1>.

Roussinos, D., & Jimoyiannis, A. (2017). Students' collaborative patterns in a wiki project: Towards a theoretical and analysis framework. *Journal of Applied Research in Higher Education 9*(1), 24–39. <https://doi.org/10.1108/JARHE-05-2016-0034>.

Semeon, N., & Mutekwe, E. (2021). Perceptions about the use of language in physical science classrooms: A discourse analysis. *South African Journal of Education 41*(1), 1–11. <http://dx.doi.org/10.15700/saje.v41n1a1781>.

Singh-Pillay, A., & Naidoo, J. (2020). Context matters: Science, technology and mathematics education lecturers' reflections on online teaching and learning during the COVID-19 pandemic. *Journal of Baltic Science Education 19*(6A), 1125–1136. <https://doi.org/10.33225/jbse/20.19.00>.

Vogel, S., & García, O. (2017). Translanguaging. In G. Noblit, & L. Moll (Eds.), *Oxford research encyclopedia of education* [online]. Oxford University Press. <https://doi.org/10.1093/acrefore/9780190264093.013.181>.

Zano, K., & Charamba, E. (2021). COVID-19-induced WhatsApp platform for EFAL learners in a multilingual context. *E-Bangi Journal of Humanities and Social Sciences*, *18*(10), 36–49.

CHAPTER 8

# Infusion of Technology in the Teaching and Learning of Mathematics in a South African University

*Neliswa Gqoli*[1]
[1]Walter Sisulu University

**ABSTRACT**
The study explored the infusion of technology into mathematics teaching and learning in one university of Oliver Reginald Tambo District in the Eastern Cape Province. To prepare students for future employment in a globally competitive and dynamic digital workplace, universities should infuse technology into their teaching and learning. The COVID-19 pandemic compelled universities to shift entirely to online teaching and learning. This shift to online teaching and learning affected lecturers and university students, especially in grasping the content in mathematics classes. The study was underpinned by Mishra and Koehler's theoretical framework, the Technological Pedagogical Content Knowledge, which explains how technology should be used pedagogically in ways appropriate to the subject being taught. This was a qualitative study that utilized a case study design. The participants for the study included five mathematics lecturers purposefully selected as information-rich participants. Face-to-face semi-structured interviews were utilized to collect data, which were analyzed using thematic analysis. The study's findings revealed that though lecturers showed a positive attitude towards using technology in their mathematics teaching, there were various challenges to online learning, especially in rural universities. Due to the above findings, the study recommends that technology professional development in mathematics be offered to lecturers in rural universities. Furthermore, the other recommendations included creating user-friendly mathematics marking and preparation systems that use technology and technological mathematical pedagogies in teacher education programs.

*Keywords:* Digital learning, infusion, learning, mathematics, rural-university, teaching

## Introduction

Technology, particularly e-learning, is being used in teaching more frequently to improve the quality of teaching and learning. Therefore, most educational programs use a variety of e-learning technologies. In many areas of the world, education departments and universities have made significant investments to increase the use of online platforms in all of their forms (e.g., e-books, simulations, text messaging, podcasting, blogs, etc.) to meet the demands of competitive markets and provide their students with a variety of learning options (Oke and Fernandes, 2020).

Heflin et al. (2017) mention that though advocates of educational change promote the need for a learning environment that prepares students to deal with changes as they occur and optimistically to help create needed changes, most university teachers continue to operate from a traditional system of existing pedagogy (Carrillo and Flores, 2020). Additionally, some universities in the Eastern Cape Province used face-to-face teaching until the rapid emergence of COVID-19, which significantly impacted educational institutions. However, mathematics teaching and learning in rural universities have been impacted by the COVID-19 pandemic. Traditional face-to-face classroom teaching of mathematics was replaced with emergency online and remote learning for both students and teachers. Universities use online learning to maintain the educational process and continuity. In addition, the students and teachers were inclined toward the acceptance of hybrid learning and their enthusiasm for online mathematics programs, especially in rural areas (Dhawan, 2020). Hence, studying how technology is infused into mathematics learning and teaching is crucial, especially in rural universities.

## Literature Review

Technology integration in teaching and learning includes formal and informal use of Information and Computer Technology by students, teachers, and others inside and outside of the classroom to improve the performance of a multidimensional information system. Tabach (2011) believes that technology integration into mathematics classes is growing. Additionally, utilizing technology transforms the teaching and learning process and produces a powerful learning environment where students engage with knowledge practically and productively. Therefore, as technology spreads, more people are interested in adopting mobile devices to support teaching and learning, especially in mathematics (Kearney and Maher, 2019). However, due to the COVID-19 pandemic's rapid spread, traditional classroom learning and teaching have been disrupted. As a result, universities have been forced to implement alternative strategies that follow the social distancing guidelines to prevent contracting or spreading the virus.

According to Engelbrecht et al. (2020), Internet development and its accessibility have dramatically changed how two-way communication can happen between students and between students and teachers. New technologies have extended the concepts of classroom and lecture, and it is not easy to distinguish between inside and outside the classroom and between study and leisure time. Borba et al. (2017) pointed out that online mathematics learning resources challenged the traditional image of the flow of mathematical knowledge from teachers to students. As mathematics educators, we must consider whether these resources are designed to foster a meaningful understanding of mathematics. The COVID-19 pandemic has allowed many students to use technology resources before using textbooks or consulting their mathematics teachers and lecturers. However, Attard (2015) and Freeman et al. (2017) think that using digital technologies in mathematics can be ineffective, distracting, or even dangerous when not integrated into the learning process in meaningful ways. Additionally, integrating technology into mathematics teaching and learning is a complex task that requires the consideration of many elements, including pedagogy, content, and student learning. Therefore, this calls for online learning environments that allow mathematics teachers to interpret practices as they currently occur in mathematics classrooms, disrupting traditional methods to implement meaningful student-student and student-teacher interactions through blended learning approaches.

The researchers Al-Ruz and Khasawneh (2011) state that the use of technology in teaching and learning is directly related to teachers' self-efficacy, aptitude, and effectiveness in infusing technology into instruction. Furthermore, teacher training programs that infuse technology are vital. Therefore, an effective infusion of technology into the teacher training curriculum would motivate mathematics teachers and boost their confidence in using technology to facilitate mathematics teaching. Silva et al. (2021) state that there are a variety of advantages to using technology in the classroom, including increased task-based learning, active learning, collaborative learning, and student autonomy. Additionally, Howard (2019) reveals that technology integration also has a relationship with several dynamic characteristics, such as effective practices, the technological features of new tools, the capacity to transform learning, and the availability of new teaching and learning paradigms. Therefore, teacher education programs can play a crucial role in formulating student teachers' views and intentions and providing pedagogical learning experiences so that future mathematics teachers can judge when it is appropriate to integrate information and communications technology (ICT) in their classes (Nikolopoulou and Gialamas, 2009). Hence, teachers should be encouraged to pedagogically experiment and explore digital practices with students and all participants involved in teaching and learning (Lafton, 2012). However, the literature mentioned the importance and advantages of infusing technology in teaching and learning. However, there is little to no information on how teachers in rural universities should integrate technology into mathematics teaching involving students. Furthermore, the pedagogies that teachers in rural universities should use in technology integration are not specified. This has piqued the researcher's curiosity about how teachers in rural universities use technology to enhance mathematics.

## Objectives

The objectives of the study are:

- To explore how technology is infused into the teaching and learning of mathematics in rural universities.
- To explore hindrances that impede technology integration in teaching mathematics in rural universities.

## Research questions

- How is technology infused into the teaching and learning of mathematics in rural universities?
- What are the hindrances impeding technology integration in teaching mathematics in rural universities?

## Theoretical Framework

The study is underpinned by the Technological Pedagogical Content Knowledge (TPACK) theory by Mishra and Koehler (2008). TPACK refers to the intersection of instructors' technological, pedagogical, and content knowledge. Mishra (2019) stated that TPACK is the fusion of instructors' PK, CK, and technology expertise (TK). In this regard, TPACK is a complex body of information influenced by a wide range of contextual factors, many of which are dynamic or evolving quickly, such as school organization, curricula, student socioeconomic backgrounds, and technology. Most importantly, teachers should understand how to combine two (TCK and TPK) or all three (TPACK). Developing TPACK requires a thorough understanding of how technology, content, and pedagogy are interconnected, and that understanding must be applied to develop suitable representations and procedures for the setting (Figure 8.1).

The study used TPACK because teachers should be able to comprehend how technology can be used pedagogically in ways that are relevant to the subject(s) being taught (Koehler et al. et al., 2014). Mertala (2016) reveals that for teachers to implement technology in their classes effectively, they need to be well-informed about digital media cultures, which are vital for children. This does not imply, however, that teachers must be fully aware of everything that students use and engage in in the digital sphere. In contrast, having some experience with the mathematics environment is helpful so that teachers can let students be the experts regarding their digital behaviors and inform teachers about these (Parry, 2013). Therefore, when integrating technology into a

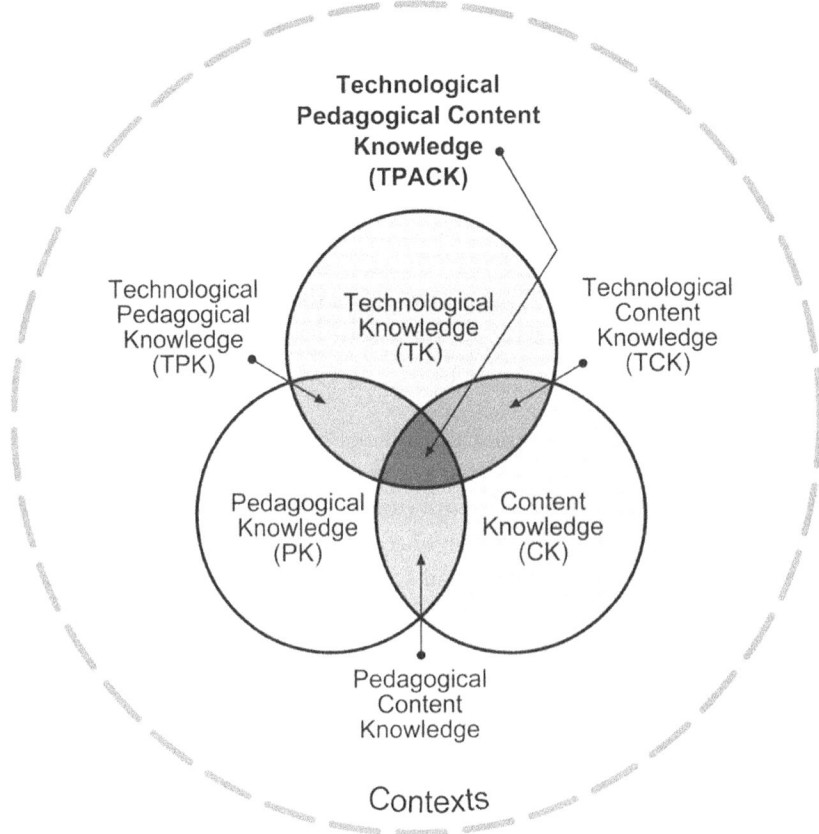

**Figure 8.1:** The technological pedagogical content knowledge model.

lesson, teachers should consider the level of technology proficiency the students need, the mathematical material needed, and the most effective ways to teach technology and mathematics.

## Methodology

The study adopted a qualitative research approach to determine the significance of a phenomenon (technology) from the participants' perspectives (Creswell, 2018).

### Research design

The case study research design, which is defined by Yin (2014) as an empirical inquiry that investigates a contemporary phenomenon (the 'case') in depth and within its real-world context, was employed. This design is used in the study because it is an empirical inquiry that investigates a contemporary phenomenon (technology integration) in depth and within its real-world context (Rashid et al., 2019), which can help explain the outcomes by matching empirically observed events with theoretically predicted events.

## Population for the study

The study's population included five lecturers in the Mathematics Department in the Faculty of Education at one university in the OR Tambo Inland District in the Eastern Cape Province.

## Sampling

Five mathematics lecturers in the Department of Mathematics in the Faculty of Education were purposefully selected as the participants for the study. The lecturers were purposefully selected as information-rich participants for the study. According to Kelly et al. (2010), purposeful sampling is used to pick respondents who are most likely to give suitable and helpful information and is a method of discovering and choosing cases that will make good use of the limited research resources available (Palinkas et al., 2015).

## Data collection instruments

The study used face-to-face, semi-structured interviews to gather information from five participants who were teaching mathematics in a rural university in the Eastern Cape Province within their life contexts.

## Data collection procedures

An ethical clearance certificate was obtained from the university under study. To ensure that the research process was ethical, the researcher saw that several ethical protocols were in place, including participants' rights, informed permission, professionalism, secrecy, protection from harm, anonymity, and participants' vulnerability. An interview guide with questions about the use of technology in mathematics instruction and learning was also provided. Data were gathered during the interviews using a tape recorder, and participants were probed for additional information.

## Data analysis procedures

The study used thematic analysis by Creswell (2018), which divides data into segments and then proceeds to assign codes, categories, and themes to those segments. According to Creswell (2018), the data were analyzed using thematic analysis in the following three processes, which are as follows:

1. Organizing the data and defining the code is the first step.
2. Creating the categories and codes is step two.
3. The third step is creating themes and sub-themes from the data.

# Findings
## Integration of technology into the teaching and learning of mathematics
Sub-theme: little knowledge of computer skills and connection problems

The findings of the study revealed that mathematics teachers were forced to use technology in their teaching due to the COVID-19 pandemic, and they used trial and error to conduct their lessons. At some stages, there were disturbances due to connection problems. These are their responses when asked about how they teach mathematics using technology:

> P1: "I was capacitated to teach online, but the training we had was not enough. For now, we teach online lessons on a trial-and-error basis as individual teachers. The school only provides Wi-Fi for both teachers and students, but there is a challenge of network and load shedding".

> P2: "In maths, you can teach nearly all the topics online. However, topics like transformation are complicated to teach online. In my experience, transformation is a hands-on topic. It needs resources to demonstrate how you translate, rotate, stretch, etc., an object. When you are online, showing reality is a bit challenging".

> P3: "As a teacher with many students in my class, I struggle to load them on my computer, and it is worse when it comes to marking. I cannot mark assignments and ended up giving students multiple choice questions with few discussion questions to save time. Even then, students complain about lack of network and low batteries caused by load shedding".

> P4: "Teaching practical work online does not work. Mostly, the teacher demonstrates, and you cannot teach effectively by just demonstrating. The students themselves need hands-on practice. They need that ability to manipulate apparatus to draw conclusions from their work".

> P5: "I was excited when we started this online activity since anything new is thrilling, but the joy quickly faded when I realised it needed a lot of work to prepare one lesson. I slept late at night for the first few weeks to gather the information and spend hours marking. While teaching face-to-face, you only need to mark the students' work with a red pen. However, when teaching online, you must decide how to display your marking; whether it is a Word document or a PDF, you must correct the diagrams. The process of marking students' assignments takes a very long time".

The results above demonstrate that teachers know the importance of online teaching and learning for student collaboration. Some participants' responses indicated the importance of closing the resource gap as universities steadily shift to twenty-first-century learning environments. However, teachers in rural universities are impacted by their lack of experience in technology and the challenges they face, especially while teaching mathematics online.

## Sub-theme: shortage of technology resources

The study's findings revealed teachers' positive dispositions towards face-to-face teaching compared to online instructional modes. This was caused by the shortage of resources, such as poor internet connections, low bandwidth, and electricity load shedding in the country, which had strained the smooth running of online mathematics classes. Following are their responses when they were asked about the hindrances that impede their online teaching of mathematics

> P1: "During online teaching, some students call to say they have lost internet connectivity or their devices have no battery and electricity. Therefore, they are not able to join the lesson. All this happens when you are trying to present a lesson."

> P2: "During online teaching, there is no classroom monitoring, and some students do not join online classes. Others just go offline if they feel they do not like the lesson. As a teacher, you just move with those who want to move with you. Also, there is limited engagement and participation in the class activities on the side of the students. Some students are shy to ask questions, and you assume they understand. However, with face-to-face, the teacher can move around the classroom, identifying those students who are doing or not doing well, and then you can assist. However, during online teaching, it is not easy to conduct remedial lessons".

## Sub-theme: minimal training on technology

Teachers' beliefs about online teaching and learning showed that some teachers lacked knowledge of how to conduct online classes because of the crash course provided. Also, teachers experienced difficulties in marking, resulting in them preferring traditional teaching methods. These are some of their responses:

> P1: "I lack knowledge of using virtual classrooms and could not use the online platforms because we had little training. I had to get my children to do it for me, but eventually, I adapted. Initially, I thought teaching on online platforms would not work. I was very negative about it, to be honest. Nevertheless, I tried it, especially when the connectivity is right".

> P4: "The professional development was not thorough because it was a crash course. I experienced challenges understanding how to operate the online platforms and that three days of training was not enough, especially for older teachers like me."

> P5: "I was initially positive about the new programme, but the challenges I experienced compelled me to favour the old system. The marking system has changed significantly and can sometimes slow the online teaching process. It is more difficult to mark and include explanations while marking than writing on paper."

## Discussion of Findings

Tabach (2011) mentioned that technology integration into mathematics classes is growing. Crompton et al. (2017) mention the need to change traditional teaching methods from reactionary approaches and acquire short-term skills to proactive programs that necessitate lifelong learning attitudes that are important as universities prepare students for the twenty-first century. However, the study's findings, categorized under the theme of integrating technology into mathematics teaching and learning, showed that lecturers in rural universities struggled to conduct online mathematics lessons during the COVID-19 pandemic. Furthermore, Koehler et al. (2014), teachers should be able to comprehend how technology can be used pedagogically in relevant ways to the subject(s) being taught.

Mertala (2016) also reveals that for teachers to implement technology in their classes effectively, they need to be well-informed about digital media cultures, which are vital for children. Nevertheless, the teachers in rural universities needed more technical training and had a crash course on online teaching, which was not enough. As a result, they used trial and error to teach mathematics. Additionally, some teachers were interested in teaching mathematics using technology but needed help with some topics that needed hands-on activities.

The study also showed that when teachers in rural universities discovered they could not teach practical work like translation online, their enthusiasm for teaching with technology diminished. Also, many found the long preparation and online marking times frustrating.

Furthermore, researchers Attard (2015) and Freeman et al. (2017) believe that using digital technologies in mathematics can be ineffective, distracting, or even dangerous when not integrated into the learning process in meaningful ways. Some teachers' responses indicated the importance of closing the resource gap as universities steadily shift to twenty-first-century learning environments. According to Mishra and Koehler (2008), developing TPACK requires a thorough understanding of how technology, content, and pedagogy are interconnected, and that understanding must be applied to develop representations and procedures suitable for the setting. However, this is lacking in teachers of rural universities due to a lack of experience in technology and the challenges they face, especially while teaching mathematics online.

The following theme is on hindrances that impede the integration of technology in teaching mathematics in rural universities. The study's findings revealed various challenges experienced by the teachers in rural universities when implementing mathematics online, including loss of connectivity caused by network problems and load-shedding. Additionally, the lack of classroom monitoring and student engagement in the class activities was a challenging factor that made the teacher assume that the students understood the lesson's content. Nonetheless, the researchers Al-Ruz and Khasawneh (2011) believe that using technology in teaching and learning is directly related to teachers' self-efficacy, aptitude, and effectiveness in infusing technology into instruction. Furthermore, teachers' beliefs about teaching using technology indicated that teachers in rural universities were of the opinion that the use of technology is necessary to develop students' skills and knowledge so that they can operate effectively, being able to create and express themselves within a digital environment.

Additionally, mathematics teachers in rural universities believe that COVID-19 has created a platform where many applications can be used to represent and discuss mathematical concepts using technology. Therefore, many online platforms also ease the communication between students and lecturers and among students. Hence, the importance of teacher education programs that can play a crucial role in formulating teachers' views and intentions, providing pedagogical learning experiences so that future teachers will be able to judge when it is appropriate to integrate ICT in their classes, is vital (Nikolopoulou and Gialamas, 2009).

## Recommendations

Based on the findings above, the study recommends that:

- Technology professional development in mathematics should be offered to teachers at rural universities.
- Universities should create user-friendly, technology-based mathematics marking and preparation systems and develop methods for monitoring and controlling mathematics test writing.
- The universities should strengthen access points and offer backups to accommodate load shedding.
- The use of technological mathematical pedagogies in teacher education programs is recommended.
- Another recommendation is that universities in rural areas should set policies demanding that teachers attend technology training programs to enhance the teaching and learning process and have technology mentors on campus to better meet the teachers' needs and questions.

## Conclusion

In conclusion, teachers in rural universities believe that technology may significantly improve mathematics education. Additionally, using technology in mathematics encourages student collaboration because it allows them to discuss and develop alternative approaches to solving mathematical problems. However, they came across difficulties when teaching online. The study suggests a number of measures to assist teachers in integrating technology into their mathematics teaching and learning, which include professional development in technology, provision of programs for mentoring and invigilation, backup power during load shedding, the availability of technological pedagogies, and policies requiring university teachers to attend technology training programs.

## Limitations

The study's focus was solely on teachers as units of analysis, and the results are only applicable to the sample of five university teachers. The results only apply to the sample of five university teachers because the study's main focus and analytical units were teachers. Generalizations to the broader population are difficult to establish because the study only included a small sample of five teachers from one university in the OR Tambo Inland District. As a result, future research should use a combination of methods to give population-level evidence.

## Bibliography

Al-Ruz, J. A., & Khasawneh, S. (2011). Jordanian pre-service teachers' and technology integration: A human resource development approach. *Journal of Educational Technology & Society 14*(4), 77–87.

Attard, C. (2015). Introducing iPads into primary mathematics classrooms: Teachers' experiences and pedagogies. In *Integrating Touch-Enabled and Mobile Devices into Contemporary Mathematics Education* (pp. 193–213). <https://doi.org/10.4018/978-1-4666-8714-1.ch009>.

Borba, M. C., Askar, P., Engelbrecht, J., Gadanidis, G., Llinares, S., & Aguilar, M. S. (2017). Digital technology in mathematics education: Research over the last decade. In *Proceedings of the 13th International Congress on Mathematical Education: ICME-13* (pp. 221–233). Springer International Publishing.

Carrillo, C., & Flores, M. A. (2020). COVID-19 and teacher education: A literature review of online teaching and learning practices. *European Journal of Teacher Education 43*(4), 466–487.

Creswell, J. W. (2018). *Research design: Qualitative, quantitative, and mixed methods approaches*. 4th ed. London: Sage.

Crompton, H., Burke, D., & Gregory, K. H. (2017). The use of mobile learning in PK-12 education: A systematic review. *Computers & Education 110*, 51–63.

Dhawan, S. (2020). Online learning: A panacea in the time of COVID-19 crisis. *Journal of Educational Technology Systems 49*(1), 5–22.

Engelbrecht, J., Llinares, S., & Borba, M. C. (2020). Transformation of the mathematics classroom with the internet. *ZDM 52*, 825–841. <https://doi.org/10.1007/s11858-020-01176-4>.

Freeman, A., Adams Becker, S., Cummins, M., Davis, A., & Hall Giesinger, C. (2017). *NMC/CoSN Horizon Report: 2017 K-12 Edition*. Retrieved from <https://www.nmc.org/publication/nmccosn-horizon-report2017-k-12-edition/>.

Heflin, H., Shewmaker, J., & Nguyen, J. (2017). Impact of mobile technology on student attitudes, engagement, and learning. *Computers & Education 107*, 91–99.

Howard, T. C. (2019). *Why Race and Culture Matter in Schools: Closing the Achievement Gap in America's Classrooms*. Teachers College Press.

Kearney, M., & Maher, D. (2019). Mobile learning in pre-service teacher education: Examining the use of professional learning networks. *Australasian Journal of Educational Technology 35*(1).

Kelly, S. E., Bourgeault, I., & Dingwall, R. (2010). Qualitative interviewing techniques and styles. *The SAGE Handbook of Qualitative Methods in Health Research 19*, 307–326.

Koehler, M. J., Mishra, P., Kereluik, K., Shin, T. S., & Graham, C. R. (2014). The technological pedagogical content knowledge framework. *Handbook of Research on Educational Communications and Technology*, 101–111.

Lafton, T. (2012). How early childhood practitioners build, shape, and construct their digital practices: The search for an analytical space. *Nordic Journal of Digital Literacy 7*(3), 172–185.

Mertala, P. (2016). Digital technologies in early childhood education–a frame analysis of preservice teachers' perceptions. *Early Child Development and Care 189*(8), 1228–1241.

Mishra, P. (2019). Considering Contextual Knowledge: The TPACK Diagram Gets an Upgrade. *Journal of Digital Learning in Teacher Education, 35*, 76–78.

Mishra, P., & Koehler, M. J. (2008, March). Introducing technological content knowledge. In *The Annual Meeting of the American Educational Research Association 1*(16).

Nieuwenhuis, J. (2016). Analysing qualitative data. In Maree, K. (Ed.). *First Steps in Research*, 2nd ed. Pretoria: Van Schaik.

Nikolopoulou, K., & Gialamas, V. (2009). Investigating pre-service early childhood teachers' views and intentions about integrating and using computers in early childhood settings: Compilation of an instrument. *Technology, Pedagogy and Education 18*(2), 201–219.

Oke, A., & Fernandes, F. A. P. (2020). Innovations in teaching and learning: Exploring the perceptions of the education sector on the fourth industrial revolution (4IR). *Journal of Open Innovation: Technology, Market, and Complexity 6*(2), 31.

Palinkas, L. A., Horwitz, S. M., Green, C. A., Wisdom, J. P., Duan, N., & Hoagwood, K. (2015). Purposeful sampling for qualitative data collection and analysis in mixed method implementation research. *Administration and Policy in Mental Health and Mental Health Services Research 42*(5), 533–544.

Parry, J. P. (2013). *Caste and Kinship in Kangra*. Routledge.

Rashid, Y., Rashid, A., Warraich, M. A., Sabir, S. S., & Waseem, A. (2019). Case study method: A step-by-step guide for business researchers. *International Journal of Qualitative Methods 18*, 1609406919862424.

Silva, O., Sousa, Á., & Nunes, J. (2021, February). Factors that affect student's attitudes towards statistics and technology and their interrelationships. In *International Conference on Information Technology & Systems* (pp. 157–166). Cham: Springer.

Tabach, M. (2011). A mathematics teacher's practice in a technological environment: A case study analysis using two complementary theories. *Technology, Knowledge and Learning 16*, 247–265.

Yin, R. K. (2014). *Case Study Research: Design and Methods*. New York: Sage.

CHAPTER 9

# STEM Teaching and Learning in Early Childhood Classrooms During and Post the COVID-19 Pandemic in Zimbabwe

Agnes Pakombwele[1]
[1]Baisago University, Botswana

**ABSTRACT**
The outbreak of the coronavirus disease (COVID-19) all over the world challenged the teachers' efforts and practices, creating the need to change and adapt to new pedagogies. The COVID-19 pandemic began when some countries, such as Zimbabwe, were crafting policies to implement science, technology, engineering, and mathematics (STEM) education from early childhood development (ECD) to higher education. The competency-based curriculum framework in Zimbabwe underpinned STEM education as an approach to developing global competencies (twenty-first-century skills) in ECD learners. This study explored ECD teachers' practices and experiences in STEM teaching and learning during and after the COVID-19 pandemic lockdowns. The study adopted an interpretive paradigm and multiple case study design involving three schools, where three ECD teachers (one from each school) were purposively selected. Data were solicited through semi-structured interviews, document analysis, and classroom observations. It can arguably be acknowledged that the COVID-19 lockdowns caused significant disruption to STEM education as learners and teachers suffered physical disconnection from schools. The barriers that teachers and learners faced encompassed poor adherence to COVID-19 restrictions, lack of proper assessment of activities done at home, inadequate resources, limited teaching time, and lack of training. The study recommends implementing long-term educational policies and practices to enable educators to prepare for the next pandemic or other societal disruptions. Early childhood educators need the necessary skills through training to improve their STEM expertise. Such training would mold teachers' understanding of different teaching approaches in the advent of unprepared outbreaks.

*Keywords:* COVID-19, digital tools, digital platforms, ECD teachers, early childhood, lockdown, pandemic, STEM education

## Introduction

The government of Zimbabwe's major thrust on teaching science, technology, engineering, and mathematics (STEM) subjects from early childhood development (ECD) sought to industrialize the economy and create much-needed employment in the country (Mutseekwa, 2021). In 2015, the Ministry of Primary and Secondary Education (MoPSE) introduced a competency-based curriculum (CBC 2015) which emphasized STEM/STEAM education for the subsequent building of global competencies in every learner from ECD to advanced level (Lower and Upper 6) (Curriculum Framework for Primary and Secondary Education, 2015).

The competency-based curriculum is divided into three terms: January-March (first term), May–July (second term), and September–November (third term). In March 2020, Zimbabwe joined the rest of the world in a global crisis involving the coronavirus disease 2019 (COVID-19) outbreak caused by the virus known as severe acute respiratory syndrome coronavirus (World Bank, 2008). National and state leaders mandated quarantine measures, social distancing, and wearing mask orders (Moyo, 2020). In Zimbabwe, schools nationwide closed their buildings based on recommendations and guidance from the Ministry of Health and Child Welfare. The MoPSE directed schools to use remote methods of instruction after the school closure was announced (Mukomana, 2019; Moyo, 2020). The advent of COVID-19 changed the education system's implementation and practices of stakeholders such as teachers, learners, and parents. Learners in rural and urban areas were physically and psychologically affected by the closure of schools and entire cities (Pascal et al., 2020; Szente, 2020; Moyo, 2020). During this period, there were two options: suspending ongoing education and closing all schools to prepare for online education for an uncertain period. The second option was to shift traditional, face-to-face education to an online mode of delivery (Ozturuk and Gangal, 2022). The two options allowed teachers to utilize digital technologies in education (Ahlam, 2022). Given this scenario, teachers embraced digital technology to implement online education using various digital platforms and tools. Against this backdrop, the chapter explored the digital tools and platforms that ECD teachers used in teaching STEM during and immediately after the COVID-19 lockdowns.

## Literature Review

The literature was guided by the following subtopics: teaching early childhood learners during the COVID-19 pandemic, STEM teaching in early childhood, and the research's theoretical framework. The aim was to explore the different methodologies employed by ECD teachers during COVID-19.

## Teaching Early Childhood Learners During the COVID-19 Pandemic Period

The World Health Organization declared the outbreak a global pandemic in March 2020 (WHO, 2020). The pandemic forced educational institutions in many countries worldwide to close (Nikolopoulou, 2022). The educational process continued online in higher education institutions and some primary and secondary schools in Zimbabwe. To compensate for the time lost, some teachers sought to utilize digital technologies in education (Nikolopoulou, 2022; WHO, 2020). Most teachers had no option but to use various digital platforms and tools to deliver content. Empirical evidence suggests that several types of research were conducted worldwide to investigate how preschool education was implemented during the pandemic. The research studies included the nature of activities done by learners, the challenges incurred, and the sustainable measures taken to sustain preschool education.

In a study on preschool education in Turkey during the COVID-19 pandemic, findings by Yildirim (2021) revealed that most teachers reported that the pandemic negatively affected students (in terms of their emotional, cognitive, and psychomotor skills and teacher-student interaction). The activities performed were art (Turkish) language, science, drama, music, mathematics, and games, while the skills-concepts the teachers wanted their students to develop included collaboration, flexibility, as well as learning hygiene, numbers, and shape-related concepts (Yildirim, 2021; Nikolopoulou, 2022). The study concluded that, despite facing major challenges regarding internet connection, higher parental engagement in children's education was reported. In addition, communication among teachers, parents, and students was mostly through phone calls and WhatsApp.

Similarly, in another study on the effects of the COVID-19 pandemic on preschool education in Turkey, Duran (2021) revealed that the activities most preferred by preschool teachers were games and art activities about hygiene and communication with parents over internet platforms (WhatsApp, Zoom, and Skype) or telephone calls. In the same study, teachers expressed some negative feelings (e.g., anxiety, worry, fear), while a positive effect of the pandemic was parents' participation in children's educational activities. In the same country, Alan (2021) also conducted a study on distance education during the COVID-19 pandemic and identified the needs of early childhood teachers regarding online education during the pandemic. The study found that teachers ought to improve their technological competencies, have more interactive resources, and have a user-friendly educational platform (to provide educational activities and facilitate communication.

Hu et al. (2021) researched technology integration for young children during the COVID-19 pandemic. They provided evidence of how preschool teachers applied online resources (e.g., via digital-mediated learning platforms) to young Hong Kong children. Some barriers included teachers facing difficulties engaging their learners' online and inadequate parental support for learning activities. The study further reported that teachers who perceived greater children's engagement or support from parents were more likely to use online teaching in the future. More interactive online teaching preparation was suggested.

In another study, Timmons et al. (2021) examined the implementation and impact of online teaching in early primary contexts (kindergarten, grades 1 and 2) during the pandemic in Canada. The challenges perceived by teachers in the study included equity issues (concerns with access to technology and support at home), the social, emotional, and academic effects of online teaching for children, and its effects on parents. In their recommendations, the researchers suggested improving online teaching-learning in the post-COVID-19 era.

In the USA, Steed and Leech (2021) reported that the most utilized learning activities were singing songs and reading stories online. Teachers expressed concerns such as missing in-person interactions with children, concerns about children with special needs, inadequate resources, and a lack of guidance from administrators (about online education). The role of families was crucial since teachers relied on them to implement children's activities at home. In a similar study, Atiles et al. (2021) explored the experiences of teachers of 3–6-year-old children in the USA and some Latin American countries during the pandemic. The findings revealed that various platforms were used to communicate with parents, while half of the sample faced internet connectivity problems. Teachers expressed concerns for children's and families' mental health; they questioned the suitability of online learning for young children and reported a lack of preparation/training for online teaching. The above literature suggests a dearth of literature on ECD teachers' STEM teaching experiences during the COVID-19 pandemic. Given the above literature, it becomes imperative to explore ECD teachers' practices during and post-COVID era, specifically focusing on STEM teaching and learning in Zimbabwean classrooms.

## STEM Teaching in Early Childhood

The 2030 Agenda for Sustainable Development has paved the way for pre-schools to facilitate the development of competencies for all young children. STEM education is a valuable model for developing these competencies (Aguilera and Ortiz-Revilla, 2021; Johnston et al., 2022; Gordy et al., 2022). The model lays down a solid foundation of learning in early childhood that stays with children throughout their lives and helps them perform better in every sphere of their lives. ECD is a crucial period of learning and development, and STEM education equips young learners with the required competence and knowledge for the future workforce in the twenty-first century.

STEM education calls for classroom practices that are supportive and transformative, meeting the needs of twenty-first-century learners. STEM learning in early childhood is the fundamental area in the twenty-first century that prepares learners to solve societal problems (Mutseekwa, 2021). Empirical literature reveals that young children need to be equipped with a range of problem-solving and creative thinking skills to support their societies in different communities later in life (Lange et al., 2022; MacDonald et al., 2022). During this twenty-first era, learners need an education that develops successful learning, living, and working, or education that can be applied to real-world context (Lange et al., 2022).

Furthermore, STEM education is a human capital development strategy that is useful for the industrialization and modernization of Zimbabwe in order to achieve its goals and objectives (Moyo, 2020; Mutseekwa, 2021). Thus, Zimbabwe's vision to become a middle-income economy by 2030 heavily lies on ECD) teachers embracing STEM education during science teaching. STEM teaching prepares learners for the fourth industrial revolution (Kim, 2020) and caters to their total development through the practical application of lifelong skills (Perales and Arostegui, 2021). STEM teaching in ECD becomes relevant because it equips learners with knowledge, skills, and traits that can effectively function as citizens, workers, and leaders in the twenty-first-century workplace (Aguilera and Ortiz-Revilla, 2021). To this end, effective investment in STEM education is critical to achieving the industrialization and modernization of Zimbabwe (Mutseekwa, 2021). Improving the teaching of STEM in early childhood classrooms has become critical during the COVID-19 era, considering the value of the STEM curriculum in the country (Mukomana, 2019; Mutseekwa, 2021).

## Theoretical Framework

The study adopted Schön's notion of reflection on action and reflection in action (Schön, 1983). This theory was employed to explore teachers' reflections on their online classroom experiences of teaching ECD learners during and after the COVID-19 pandemic lockdowns. According to Schön (1983), reflection-in-action is the process that allows professionals to reshape the situation or activity on which they are working while it is unfolding. It is generally associated with the experience of surprise (Bolton, 2014). The COVID-19 surprised the entire education fraternity, including ECD teachers. The teachers had no option but to redirect their teaching practices to suit the prevailing situation. Schon suggests that, by reflecting in action, professionals reflect on unexpected experiences and conduct "experiments" to generate a new understanding of the experience and a change in the situation. Reflecting in action refers to situations such as thinking on your feet, acting straight away, and thinking about what to do next (Bolton, 2014). In this context, reflection in action refers to learning by experience, where teachers encounter a new perspective on practice, such as teaching and learning STEM online.

On the other hand, reflection-on-action involves reflecting on an experience, situation, or phenomenon after it has occurred (Schön, 1983). When professionals reflect, they explore what happened in that situation, why they acted as they did, whether they would have acted differently, and so on. Soon after the pandemic, the ECD teachers were expected to reflect on their practices during the pandemic and come up with better solutions as a way of equipping themselves for other eventualities in the future. Reflection-on-action is often associated with reflective writing, in which professionals reflect on their experiences and examine alternative ways to improve their practice (Loughran, 2002). Similarly, reflecting on action means thinking about what a teacher would do differently next time after taking time to process (Bolton, 2014). The study findings show that ECD teachers took time to reflect on their practices and implemented relevant teaching methodologies during the COVID-19 lockdowns.

The theory further explains reflection as a complex process during which teachers re-examine their teaching practice in terms of pedagogy, content knowledge, skills, student factors, and contextual factors that impact their teaching and learning environment (Schön, 1983). In this study, ECD teachers were compelled to reflect and realign their teaching practices to match the methodology employed and skills. ECD children in the study were learning from home, which could have impacted the teachers' acquired knowledge and routine lesson delivery skills. Reflection is a process whereby teachers are cognizant of their actions at all times at three levels, namely professional, societal, and personal, to improve their practice (Hullinger et al., 2019). According to Loughran (2002), reflection involves thinking about one's teaching practice and establishing how it might be altered to enhance and promote learning (Hullinger, 2019). Loughran (2002) elaborates that, in the reflection process, teachers think deeply about their teaching at three levels, namely professional, societal, and personal, to improve both teaching and learning. As discussed earlier, the literature is replete with studies on barriers to online blended learning under normal circumstances, including studies on online distance learning (Hu et al., 2021; Timmons et al., 2021; Steed and Leech, 2021; Nikolopoulou, 2022; Yildirim, 2021). There is a lack of studies exploring online STEM teaching and learning of early childhood learners during pandemics and conditions of lockdown.

## Methodology

This study adopted a qualitative approach to answer the research questions. According to Mukherji and Albon (2015), qualitative methods are best utilized when the meaning-making perspectives of actors in a particular event are being studied. This research aimed to get insights, perspectives, and in-depth information about teachers conducting lessons during the COVID-19 lockdown period.

Three schools in the Harare-Chitungwiza metropolitan province were purposively sampled to participate in the study for convenient access to site participants and observational data. Pseudonyms were used for both school names and participant identities. Of the three, two are urban schools (Musasa and Mupani), and one, Muhacha, is in the peri-urban. Table 9.1 shows the demographic data of the participants.

Table 9.1: Demographic data of participants

| Name of School | Name of Teacher | Teachers' Qualifications | Teachers' Experience | Level Taught |
|---|---|---|---|---|
| Musasa Primary | Teacher1:T1 | Diploma in Education | 20 years | ECD A |
| Mupani Primary | Teacher2:T2 | Diploma in Education Bachelor of Education | 16 years | ECD B |
| Muhacha Primary | Teacher3:T3 | Diploma in Education | 9 years | ECD B |

Musasa and Mupani schools have two deputy heads each and practice hot sitting, indicating they are big schools. Muhacha School is small, which could be due to its location.

Three teachers, one from each school, participated in the study. Regarding gender, the participants were two female teachers, one from Musasa, the other from Mupani, and one male teacher from Muhacha. The participants were qualified and experienced in teaching their ECD classes. Consent for participation was documented by signing an informed consent form. All participants volunteered for the study. There was no monetary benefit for their participation in the interviews. Semi-structured interviews lasted over an hour and were conducted three times for each participant.

## Data collection methods

The data collection spanned from March 2020 to April 2021. Semi-structured interviews and classroom observations were conducted over Zoom. The use of two data sources helped to triangulate research findings.

## Interviews

In this study, interviews were the primary source of data collection due to the limited access to research sites and the restriction on movement during the pandemic. Interviews were conducted to gain insight into the participants' constructions and to verification of STEM instructions during the COVID-19 pandemic. Each participant was interviewed twice (once for data collection and once for data validation) for about 90 min. The first interview followed the interview protocol. The second was a follow-up interview to validate the interview transcription and review the themes discovered during the coding process. A semi-structured interview format allowed the researcher to scaffold questions around the research topics while allowing for discussion of emerging questions and issues. The questions were descriptive, allowing the participants to describe their meaning-making experiences when teaching STEM during the COVID-19 pandemic. Interviews were audio-recorded to ensure completeness and to provide an opportunity to review later. When possible, the interview was transcribed into an MS Word document within 24 h for data analysis. To ensure security, the researcher saved all recordings and transcribed data on a password-protected personal computer. After transcription, all audio files were deleted to maintain confidentiality.

## Observations

The researcher observed three classrooms for this study; each observation lasted 90 min. All observations occurred in March and April of 2021. Participants were fully aware of the nature of the study and the fact that they were being observed. Classrooms were not video recorded due to privacy concerns. The researcher took field notes during the observations. According to Graue and Walsh (1998), data records deteriorate geometrically over time, so the researcher tried to jot down as many observation points as possible during and after data collection. To limit observational bias, the researcher kept a methodological journal and shared all field notes with the teachers after the completion of data collection. Comments and feedback from the participants were included in the analysis.

## Data analysis methods

Content-thematic analysis was used, and the codes for the data analysis were descriptive. Through the process of coding, patterns of responses were used to inform themes and categories generated in line with their relevance to the research questions (Creswell, 2012). Teachers' responses were thematically grouped into those related to the pre-determined themes of teachers' practices and experiences. Anonymity was confirmed to eliminate possible untruthful responses. The results are presented according to the four research questions of the study; in excerpts, the codes T1–T3 were used for teachers (T1: for Teacher 1, T2: Teacher 2, and T3 for Teacher 3). Data analysis was completed simultaneously with data collection whenever possible, and after data collection was completed. The general methodology informed the researcher of thematic coding (Mukherji and Albon, 2015) to develop categories from the data. Open coding was used to analyze interview transcripts and observations. Coding was conducted line-by-line by defining actions or events within each line of coding. This form of coding aided the researcher in focusing attention on participants' perspectives rather than the researcher's interpretations. Data analysis validation was addressed by utilizing member checking with interviewees after data collection. Once the interviews were transcribed, participants were given copies of the transcripts with the thematic coding. If there was a mistake or a need for clarification, the researcher corrected the information based on participants' feedback. Observed participants were also given copies of the field notes with thematic coding for review. All feedback was incorporated within the data to ensure validation.

## Findings

The findings and discussion were derived from the interpretative analysis of the data generated. Data from participants (ECD teachers) and observations were coded and categorized into themes that materialized (Nieuwehuis, 2016). The sub-themes that emerged were digital platforms used by ECD teachers during the COVID-19 pandemic, methods of instruction during the pandemic, and challenges faced by teachers during the affected period.

### Digital platforms used by ECD teachers during the COVID-19 pandemic lockdown

The COVID-19 pandemic lockdown left ECD teachers with no option but to embark on online teaching and learning. Whilst the three teachers agreed that they used the WhatsApp platform to teach STEM subjects, the study recorded that they used it differently. According to the participants, the platform was affordable to teachers, parents, and guardians. However, for the teachers, the lessons were determined by the availability of personal resources to buy data. The initial step was to create a WhatsApp group with all parents and guardians. All the teachers reported using their money to purchase data and their cell phones to communicate with learners through their parents.

Teacher 1 from Musasa Primary School explained that she would send messages on the WhatsApp platform with science content and activities that she wanted the learners to do. For example, the teacher would send a message like:

*Name and identify the simple machines used in the local environment.*

In this activity, parents at home assisted learners in doing the activity. As feedback, parents sent the responses back to the teacher. In addition, she took pictures from the learners' books and sent them through WhatsApp for learners to do picture reading. For example, pictures of domestic and wild animals from the learners' textbooks were taken and sent as attachments on the platform. Learners were then asked to group the animals.

Teacher 2 from Mupani School echoed that the school authorities were not supporting them with data, so she resorted to using the WhatsApp platform. Besides sending WhatsApp messages, she would record herself teaching for 10 min and send the recordings with activities to be done. Parents would take their time to provide feedback depending on whether they could afford the data. To add variety to her lessons, she would record a video of herself demonstrating how to measure the volume of water using different containers. After recording, she would send the video to parents with instructions. Though videos were inspiring to learners, they were not popular because most parents could not afford the data to download them.

On the other hand, Teacher 3 from Muhacha School used text messages to communicate. One of the mathematics text messages read:

> *Say numbers from 1 to 10. Count objects from 1 to 10. The objects can include animals, stones, sticks or bottle tops.*

The text messages to parents contained instructions and information about the work to be done. The school was in a peri-urban area where most parents did not have smartphones. The text messages were affordable to most parents, though getting feedback from the learners was difficult. The teacher complained that evaluating whether the learners would have counted the objects took more work.

## Methods of instruction during the COVID-19 lockdown

Teachers reported that they were using various methods to teach online lessons. During an interview, Teacher 2 said:

> *Teaching online was a unique experience for me. I had to think of creative methods to attract learners' attention because young children want funny things when learning so they do not get bored. So, I would ask someone to take a short video whilst I demonstrate something and send the video to parents.*

The teacher further explained that she took a short video to allow learners to see what happens if water is added to substances like salt, sugar, soda, mealie-meal, and flour. From the teachers' view, though videos consumed more data, they added variety to the lessons. In addition, parents reported that learners enjoyed this and participated during the lessons.

Teacher 1 also claimed she used puzzles and games to teach mathematics and science. She would download a video (from the internet) of young children playing a game or puzzle. She would then send the game/puzzle to the learners. The message from the teacher's cell phone read:

*Download and watch the video. Play the game with your friends.*

Upon receiving the game/puzzle, the learners were supposed to play the same puzzle following the video. Parents reported that these were fascinating lessons that their children enjoyed because they engaged the learners' attention.

Teacher 3 from the peri-urban school explained that he would record himself teaching a technology lesson. In that audio, there were activities that learners would perform. In another example, the teacher recorded himself telling a story about the weather. The following brief excerpt is the introduction to the story from the teacher's audio:

*One day, the hare visited his uncle Baboon, who lived in another village. When the hare started the journey, it was sunny and very hot. Later, clouds began to build up, and it started raining. The hare had no umbrella, and there was no shelter nearby …*

The teacher would then send the audio to parents, who would download it and assist children in listening to the story and completing the exercise about the story.

## Challenges faced by teachers during the COVID-19 pandemic lockdown period

Despite the advantages of teaching online, as mentioned above, teachers face challenges. The following challenges were identified: lack of parental and administrative support, lack of hands-on activities and limited interaction, lack of skills and competencies, poor management of online learning, and limited resources.

### Lack of parental and administrative support

Teachers shared their different sentiments on the support they received from parents during the COVID-19 pandemic period. During group chats, Teacher 1 said:

*Early childhood online learning entirely depends on parental involvement. The lessons are conducted at home, and parents organise learners, monitor all the activities, and provide feedback.*

On the same note, Teacher 3 reported:

*Parents are complaining about online learning. It requires them to create time and support their children, and they complain that it disturbs their day-to-day hustle.*

Evidence reveals that parents in the peri-urban area continued their errands, neglecting their children's lessons. Several studies worldwide reported a lack of parental support as a challenge that impedes STEM teaching. Few studies conducted during the COVID-19 lockdown in 2020 found that a lack of parental support was one of the main challenges experienced (Pascal et al., 2020; Yildirim, 2021; Szente, 2020).

All three participants reported that school administrators did not support their teaching of online lessons. They wanted school administrators to play a crucial role in STEM teaching and learning by buying data and renting gadgets for online lessons during the COVID-19 pandemic.

### Lack of hands-on activities and limited interaction

ECD teachers posited that STEM subjects are practical disciplines best taught using a hands-on approach. Teacher 2 lamented:

> *Online teaching limits my lesson plan and engagement of learners. I cannot plan all the practical activities as required in the syllabus because I will not be able to monitor the learners' activities.*

In agreement, Teacher 3 said:

> Learners perform better when they move, do and touch. They learn better through play. Online learning limits learners' engagement in play.

Lack of learner engagement was a big challenge, and using WhatsApp to teach STEM resulted in limited interaction between teachers and learners. Instead, there was a great need for maximum adult supervision, which was also a problem for parents and guardians in peri-urban contexts. This led Szente (2020) and Nikolopoulou (2022) to conclude that online lessons need more interaction and are more passive than face-to-face lessons. In addition, Kim (2020) concurs that one of the most significant disadvantages of online learning is limited interaction between teachers and students.

### Lack of skills to teach STEM during the COVID-19 pandemic lockdown period

The COVID-19 period caught everyone unawares. The teachers found themselves in a situation they could not avoid. They found themselves in a situation that required them to display new skills and ideas. Teacher 3 had this to say:

> *I do not have enough skills to present data online. No one taught me how to teach using WhatsApp. I think training through some workshops or short courses is ideal if we are in the same situation in the near future.*

Teacher 2 discussed good communication skills as a prerequisite for successful online teaching and learning. She said:

> We had always communicated with parents concerning their children, but this time, it was a bit different. Teaching and learning were solely online, and parents were supposed to take a greater part in teaching and monitoring. I needed communication skills so that I could negotiate and persuade them to provide feedback. Some parents said they could not afford online lessons, but I had to convince them to purchase data for online lessons.

## Teaching and learning in the post-COVID-19 lockdown period

After the COVID-19 lockdowns, the MoPSE announced that primary schools would be opened in three phases. The first phase to open was grades 6 and 7, followed by grades 3, 4, and 5, and lastly, the infant school (Grades 1, 2, ECD A, and B classes). At Musasa Primary School, Teacher 1 explained:

> The school is already practising ' hot-sitting ' due to a lack of adequate resources like physical infrastructure. The classes are divided into two groups. For example, in my class of forty-five learners, 20 attended lessons twice the first week, and twenty-five learners attended three times the following week.

On the other hand, at Mupani Primary School, the division of classes was gender-based. Boys attended school the first week, and girls attended the following week. There was no "hot-sitting" at Muhacha School, located in a peri-urban area, but learners' learning time was cut short. Instead of ending lessons at noon, lessons for infant classes ended soon after break at 10.30 am. The arrangement allowed teachers to attend to a few learners at any given time and reduce the spread of the disease.

Teachers lamented that when learners resumed face-to-face lessons, they had forgotten most mathematical concepts like height, length, and width. They had to reteach most of the concepts that were done at home. Teacher 2 said:

> There is no evidence that the lessons were done. It is difficult to evaluate the lessons sent on a WhatsApp group platform. So, the best thing to do is to reteach the concept and evaluate the lesson.

Some of the lessons sent to parents were practical, demanding close monitoring. Participants narrated that they could not rely on parents' information because of a lack of evidence.

Most learners were not comfortable with wearing masks. After break time, some would have lost their masks. Some learners at Mupani School were observed performing tasks without masks and not maintaining social distancing. Teacher 3 said:

> Young children cannot sit alone for a long time. They like interacting and playing with others, and it is very difficult for them to maintain the same position for a long time.

Interviewed teachers narrated that homework became another teaching method. This was meant to compensate for the time lost during the COVID-19 lockdowns and inadequate learning time at school. Teachers lamented that covering all syllabus topics was difficult. Teacher 3 narrated that most learners were bringing back incomplete or undone homework, evidenced by incomplete homework samples kept in children's individual portfolios. The study observed that teachers resorted to giving learners some homework even though most learners from the peri-urban school would bring uncompleted homework to school.

The current mathematics and science infant syllabus emphasizes that learners should develop concepts through play. Teachers are encouraged to include play sessions in their lessons. Teachers, however, pointed out that given the limited time, learners did not have enough time to play. Teacher 2 said that they dismissed earlier than usual. Learners were not allowed to spend many hours at school. The study observed a temporary timetable indicating that lessons were ending at eleven o'clock in the morning instead of the usual noon midday. Because of the learners' age and characteristics, Teacher 3 agreed that it was difficult for learners to spend many hours wearing a mask and practicing social distancing.

Teacher 1 said she prepared some teaching and learning materials for parents to assist their children at home. The materials had clear instructions and guidelines for parents to follow, which helped parents facilitate their children's learning at home. During that period, Teacher 1 said it was crucial to maintain regular communication channels with parents to address questions and concerns and provide feedback.

## Discussion

Online learning is defined as the use of the internet and electronic devices, for example, tablets, smartphones, laptops, and computers, to interact with knowledge and learners using various learning materials with platforms such as Zoom, Google Classroom, Blackboard, Class Dojo, and many more (Singh-Pillay and Naidoo, 2020; Nikolopoulu, 2022). The data gathered confirms that WhatsApp was the most commonly used digital platform for teaching and learning during the COVID-19 pandemic. ECD teachers in the study used text (SMS), WhatsApp messages, and audio to teach STEM subjects. Research by Moyo (2020) also suggests that the WhatsApp platform proved less expensive and more accessible to most parents (Moyo, 2020). WhatsApp is one of the most popular apps that allows users to exchange instant messages and share ideas faster and easier at a low cost (Moyo, 2020; Maphosa et al., 2020). This study determined lesson delivery by the availability of the individual teacher's resources to buy data.

Schon's theory of reflection explains that teachers should reflect on the new teaching perspective and align their actions accordingly (Hullinger et al., 2019). Teachers in the study embraced the new methodology of teaching online (WhatsApp) and aligned their practices to the new pedagogy. According to the theory, teachers reflect on their actions, decisions, and outcomes as they unfold. The reflections made them conclude that aligning current teaching practices to the demands of the COVID-19 period was not enough without parents' support. The role of parents was essential for online teaching and learning during the early years (Nikolopoulu, 2022). Findings from the study suggest that the successful implementation of online STEM teaching and learning depended on other variables such as parents' support and data availability. The new online teaching pedagogy, ushered in schools by the COVID-19 pandemic, forced teachers to change their practices.

One of the major goals stated in the MoPSE in Zimbabwe's Curriculum Framework (2015–2022) is to provide STEM teaching and learning from ECD. Improving the methods of instruction is mandatory to attain this national agenda, given that STEM education plays a significant role in shaping the future of any nation (Mutseekwa, 2021). The study findings indicated that the advent of COVID-19 opened new avenues for teachers to engage in different teaching methods apart from the known ones. Ozturuk and Gangal (2022) contend that online learning is ideal, as it boosts teachers' competencies in digital pedagogical approaches. Teachers mainly use differentiated instruction and teaching methods to enhance young children's online learning experiences. The methods included jigsaw puzzles, stories, and games. The major aim of varying teaching strategies during online STEM teaching was to engage learners and keep them alert so they do not forget about schooling.

Teachers complained that they were using their resources to teach online. Administrative support is necessary for teachers to be confident in their abilities in the classroom (Ozturuk and Gangal, 2022). Before the COVID-19 pandemic, there were numerous reports of serious physical, financial, and human resource shortages in implementing the initiative to teach and train in STEM (Dube, 2018; Mutseekwa, 2021). In Zimbabwe, teachers faced problems in STEM teaching and learning, and the pandemic perpetuated the situation. In a similar study, Moyo (2020) reported that virtual field experiences could not be attained as the Zimbabwean context did not have sufficient internet coverage to allow remote teaching and learning in schools.

Albrahim (2020) identifies a set of skills needed to teach online courses: pedagogical, content, technological, social, and communication. This study also suggested that these skills were relevant in delivering quality, age-appropriate content. Data indicated that teachers lacked the skills to deliver content online. Teaching young children online differs from teaching adults because young children cannot read instructions independently. In addition, they cannot concentrate for a long time, so they need adults, such as parents or guardians, to monitor them continuously during activities. Due to the reasons above, teachers ought to be equipped with the necessary competencies to manage young children. However, Fiock (2020) contends that online learning is ideal, as it boosts teachers' competencies and readiness to implement STEM digital pedagogical approaches.

## Conclusion

The WhatsApp platform became a popular digital platform for teachers using smartphones to teach STEM during lockdowns during the COVID-19 pandemic period. The platform was affordable, cheap, and accessible to both parents and teachers. Teachers in the study reacted to the closure of schools during the pandemic by redesigning their teaching practices to suit the prevailing situation. Given the value of STEM to ECD, teachers acted accordingly so that learners would not lose out.

Quality online STEM teaching and learning depend on a collaborative, healthy working relationship among all educational stakeholders. Such stakeholders include parents, guardians, teachers, school administrators, universities, and teacher training colleges. Parents played a big role in organizing their children and making them available for the lessons. Participants in the study acknowledged active parental involvement in young children's online learning. School administrators should provide adequate and appropriate infrastructure and resources so that teachers do not rely on the WhatsApp platform only for teaching and learning STEM. The universities and teacher training colleges must prepare future educators to teach in virtual environments by introducing courses that equip educators. Teacher education programs must expand their current practices and focus on preparing trainee teachers to teach online with pedagogical skills relevant to online settings (Bentley, 2020).

Despite the challenges identified by teachers, online teaching and learning a new phenomenon where both learners and teachers learnt new skills. Teachers should reflect upon all the practices they learned during the COVID era. The reflections could assist them in designing better collaborative measures in the future. According to the theory of reflection-on-action, professionals ought to reflect on their experiences and examine ways to improve their practices (Loughran, 2002). Upon reflection on action, teachers may realize that they lack certain competencies and skills to execute their duties online. The study, therefore, recommends that school administrators allow teachers to enroll in professional development courses related to online teaching with higher learning institutions to enhance their effective teaching abilities.

## Bibliography

Ahlam Mohammed Al-Abdullatif. (2022). Towards digitilization in early childhood education: Pre-service teachers' acceptance of using digital storytelling, comics, and infographics in Saudi Arabia. *Education Sciences 12*(10), 702. <https://doi.org/10.3390/educsci12100702>.

Aguilera, D., & Ortiz-Revilla J. (2021). STEM vs. STEAM education and student creativity: A systematic literature review. *Education Sciences 11*(7), 331–340.

Alan, Ü. (2021). Distance education during the COVID-19 pandemic in Turkey: Identifying the needs of early childhood educators. *Early Childhood Education Journal 49*, 987–994.

Albrahim, F. A. (2020). Online teaching skills and competencies. *Turkish Online Journal of Educational Technology-TOJET 19*, 9–20.

Atiles, J. T., Almodóvar, M., Chavarría Vargas, A., Dias, M. J. A., & Zúñiga León, I. M. (2021). International responses to COVID-19: Challenges faced by early childhood professionals. *European Early Childhood Education Research Journal 29*, 66–78.

Bentley, L. (2022). A snapshot of science education during COVID-19 in the spring of 2021. *The Qualitative Report 27*(10), 2208–2229. <https://doi.org/10.46743/2160-3715/2022.5486>.

Bolton, G. (2014). *Reflective practice: Writing and professional development* (4th ed.). Los Angeles: SAGE.

Creswell, J. W. (2012). *Educational Research: Planning, Conducting, and Evaluating Quantitative and Qualitative Research* (4th ed.). Pearson: Boston.

Dube, S. P. (2018). An investigation of the science, technology, engineering, and mathematics (STEM) initiative in the Zimbabwean education system. *International Journal for Cross-Disciplinary Subjects in Education* (IJCDSE) 9(1), 76–98.

Duran, M. (2021). The effects of COVID-19 pandemic on preschool education. *International Journal of Educational Methodology 7*, 249–260.

Gordy, X. Z., Sparkmon, W., Imeri, H., Notebaert, A., Barnard, M., Compretta, C., & Rockhold, R. W. (2021). Science teaching excites medical interest: A qualitative inquiry of science education during the 2020 COVID-19 pandemic. *Education Sciences 11*(4), 148. <https://doi.org/10.3390/educsci11040148>.

Graue, E. M., & Walsh, D. J. (1998). *Studying Children in Context: Theories, Methods, and Ethics*. Thousand Oaks, CA: Sage.

Hu, X., Chiu, M. M., Leung, W. M. V., & Yelland, N. (2021). Technology integration for young children during COVID-19: Towards future online teaching. *British Journal of Educational Technology 52*, 1513–1537.

Hullinger, A. M., DiGirilamo, J., & Tkach, T. (2019). Reflective practice for coaches and clients: An integrated model for learning. *Philosophy of Coaching 4*(2), 5–34.

Johnston, K., Kervin, L., & Wyeth, P. (2022). STEM, STEAM and maker-spaces in early childhood: A scoping review. *Sustainability 14*(20), 13533. <https://doi.org/10.3390/su142013533>.

Kim, J. (2020). Learning and teaching online during COVID-19: Experiences of student teachers in an early childhood education practicum. *International Journal of Early Childhood 52*, 145–158. <https://doi.org/10.1007/s13158-020-00272-6>.

Lange, A. A., Robertson, L., Tian, Q., Nivens, R., & Price, J. (2022). The effects of an early childhood-elementary teacher preparation program in STEM on pre-service teachers. *Eurasia Journal of Mathematics, Science and Technology Education 18*(12). <https://doi.org/10.29333/ejmste/12698>.

Loughran, J. (2002). Effective reflective practice in search of meaning in learning about teaching. *Journal of Teacher Education 53*(1), 33–43. <https://doi.org/10.1177/0022487102053001004>.

MacDonald, B. L., Colby Tofel-Grehl, & Searle, K. A. (2022). Play, problem-solving, STEM conceptions, and efficacy in STEM: An introduction to the STEM in early childhood education special issue. *Education Sciences 12*(5), 352. <https://doi.org/10.3390/educsci12050352>.

Maphosa, V., Dube, B., & Thuthukile, J. (2020). A UTAUT evaluation of WhatsApp as a tool for lecture delivery during the COVID-19 lockdown at a Zimbabwean university. *International Journal of Higher Education 9*(5). 84–94.

Moyo, N. (2020). COVID-19 and the future of practicum in teacher education in Zimbabwe: Rethinking the 'new normal' in quality assurance for teacher certification. *Journal of Education for Teaching 46*(4), 536–545. <https://doi.org/10.1080/02607476.2020.1802702>.

Mukherji, P., & Albon, D. (2015). *Research Methods in Early Childhood: An Introductory Guide*. London: Sage.

Mukomana, S. (2019). Teaching challenges affecting stem resuscitation in Zimbabwe. *European Journal of Education Studies 5*(11), 321–332.

Mutseekwa, C. (2021). STEM practices in science teacher education curriculum: Perspectives from two secondary school teachers' colleges in Zimbabwe. *Journal of Research in Science, Mathematics and Technology Education 4*(2), 75–92. <https://doi.org/10.31756/jrsmte.422>.

Nieuwenhuis, J. (2016). Qualitative research designs and data gathering techniques. In K. Maree (Ed.), *First Steps in Research* (2nd ed., pp. 72–103). Van Schaik Publishers.

Nikolopoulou, K. (2022). Online education in early primary years: Teachers' practices and experiences during the COVID-19 pandemic. *Educational Science 12*, 76. <https:// doi.org/10.3390/educsci12020076>.

Ozturuk, Y., & Gangal, M. (2022). Early childhood student teaching practicum in Turkey through emergency remote teaching during the COVID-19 pandemic. *Problems of Education in the 21st Century 80*(3), 438–455.

Pascal, C., Bertram, T., Cullinane, C., & Holt-White, E. (2020). COVID-19 and social mobility impact brief # 4. *Early Years*. Sutton Trust.

Perales, F. J., & Arostegui, J. L. (2021). The STEAM approach: Implementation and educational, social and economic consequences. *Arts Education Policy Review*. <https:// doi.org/10.1080/10632913.2021.1974997>.

Rasmitadila, A., Rachmadtullah, R., Samsudin, A., Syaodih, E., Nurtanto, M., & Tambunan, A. R. (2020). The perceptions of primary school teachers of online learning during the COVID-19 pandemic period: A case study in Indonesia. *Journal of Ethnic and Cultural Studies 7*(2), 90–109. <https://doi.org/10.29333/ejecs/388>.

Schon, A. D. (1983). *The Reflective Practitioner: How Professionals Think in Action*. London: Temple Smith.

Singh-Pillay, A., & Naidoo, J. (2020). Context matters Science, technology and mathematics education lecturers' reflections on online teaching and learning during the COVID-19 pandemic. *Journal of Baltic Science Education 19*(6), 1125–1136. <https://doi.org/10.33225/jbse/20.19.1125>.

Steed, E. A., & Leech, N. (2021). Shifting to remote learning during COVID-19: Differences for early childhood and early childhood special education teachers. *Early Childhood Educational Journal 49*, 789–798.

Szente, J. (2020). Live virtual sessions with toddlers and pre-schoolers amid COVID-19: Implications for early childhood teacher education. *Journal of Technology and Teacher Education 28*(2), 373–380.

Tefanidou, C., & Mandrikas, A. (2023). Science teaching at a distance in Greece: Students' views. *Education Sciences 13*(4), 395. <https://doi.org/10.3390/educsci13040395>.

Timmons, K., Cooper, A., Bozek, E., & Braund, H. (2021). The impacts of COVID-19 on early childhood education: Capturing the unique challenges associated with remote teaching and learning in K-2. *Early Childhood Education Journal 49*, 887–901.

World Bank. (2008). African human development series. Working Paper No. 128. Washington DC: The World Bank.

World Health Organization (2020, March 11). *WHO Director-General's Opening Remarks at the Media Briefing on COVID-19*. World Health Organization, March 11, 2020.

Yıldırım, B. (2021). Preschool education in Turkey during the Covid-19 pandemic: A phenomenological study. *Early Childhood Educational Journal 49*, 947–963.

CHAPTER 10

# STEM Education in the New Normal: Teacher Educators' Experience on the Use of Digital and Face-to-Face Pedagogy

*Esther Kibga[1] and Fredrick Mtenzi[1]*
[1]Institute of Education Development, The Aga Khan University, East Africa

**ABSTRACT**
The COVID-19 pandemic impacted a multitude of restrictions, including the closure of learning institutions to minimize mobility and physical classroom contact. The consequences were on the instructional strategies used by science, technology, engineering, and mathematics (STEM) teacher educators to facilitate learning. During this period, learning was delivered entirely through online platforms to respond to the closure of learning institutions. Notwithstanding the challenges that could interfere with formal learning through physical contact between the teacher educators and the STEM student teachers, teacher educators ought to transform the teaching didactics and instructional strategies to suit the diverse learning environment, including the abrupt changes that may be the outcome of outbreaks like the COVID-19 pandemic. This work focuses on 22 teacher educators' experiences acquired during the COVID-19 pandemic outbreak and their way forward during the new normal after the pandemic. This study employed a qualitative approach to interview research participants purposively selected through semi-structured focus group discussions. The findings indicate that teacher educators are ready for the transformations to meet the diverse changes in the learning environment. The teacher educators proposed training and improving the learning facilities to accommodate the myriad changes, including digital and face-to-face pedagogy in teaching and learning. Therefore, the study recommends that teacher professional learning programs be planned for educators to enhance their knowledge, skills, and competencies to deal with possible changes in the education system, specifically teacher education.

*Keywords:* COVID-19 pandemic, digital pedagogy, face-to-face learning, new routine, teacher educators' experiences, teacher professional learning

## Introduction

There is an ongoing debate on the suitable means for teaching and learning science, technology, engineering, and mathematics (STEM) in teacher education colleges. Whether the traditional face-to-face, that is, the use of explanation followed by procedures, hands-on, and corrections (Sullivan et al., 2020), can be replaced by or blended with the use of technology and digital pedagogy (Allen and Trinick, 2021). However, a few years after the COVID-19 pandemic, systems worldwide are forced to adapt to the "new normal" (Corpuz, 2021). It is impossible to reverse back to the old normal before the pandemic completely. All aspects of life, including personal, social, spiritual, cultural, and economic, were forced to adjust in response to the consequences of COVID-19 (Godber and Atkins, 2021). Meanwhile, individuals ought to constantly adapt to how the "new normal" has transformed the essential aspects of life, including teaching and learning.

## Literature Review

The lockdown due to COVID-19 affected approximately 15 billion learners worldwide (Allen and Trinick, 2021; Foster et al., 2022). Teachers were forced to improvise using online teaching and other means like radio and television that could help learning continue instead of the usual face-to-face way of teaching (Brunetto et al., 2022; Hamilton et al., 2020). The abrupt change became a challenge for teachers and learners to get used to the adjustment. Teachers felt pressure due to the imposed change, like a push to new norms and customs (Allen and Trinick, 2021). Such change could have also resulted in resistance due to teachers' lack of readiness and emerging beliefs as barriers to change (Foster et al., 2022).

Furthermore, the change brought about by the COVID-19 lockdown differed substantially in many ways because the change came without prior notice, which would have called for curriculum revision (Pokhrel and Chhetri, 2021). Allen and Trinick (2021) argued that during the COVID-19 lockdown, there were no prescriptions about what should be taught and how to teach (Videla et al., 2022), which usually involves teachers' lesson preparation (Godber and Atkins, 2021). Besides, teachers were left free to explore new teaching methods with little quality assurance measures from the institutions and stakeholders (Kollosche and Meyerhöfer, 2021). The resulting change, therefore, varied based on the readiness and skillfulness of teachers and learners in integrating technology into teaching and learning.

Moreover, for developing countries like Tanzania, the digital divide issues were viewed through a high lens, especially between rural and urban settlements. The digital divide was also evident between private and public learning institutions. For instance, TCRA (2022) highlighted that the estimated broadband is based on subscriptions, internet subscriptions, satellite (USB/dongles) and mobile broadband, broadband through terrestrial cable modem, fixed subscriptions to the public DSL, and wireless. Internet use was noted to be 29,913,513 users up to March 2022. Internet use (Data Traffic in Petabyte, 1 Petabyte $=1024^3$ Megabytes was seen to be 170 Megabytes in March 2022, which increased from 140 Megabytes in September 2021. Between 2016 and 2021, the number of estimated internet users, as seen in Table 10.1, whereby despite increased in some percentage each year, still, approximately 50% of Tanzanians do not use the Internet.

Table 10.1: Number of estimated Internet users and percentage

| Year | 2016 | 2017 | 2018 | 2019 | 2020 | 2021 |
| --- | --- | --- | --- | --- | --- | --- |
| Internet users | 19,862,525 | 22,995,109 | 23,142,960 | 25,794,560 | 28,470,506 | 29,858,759 |
| Percentages | 40% | 45% | 43% | 46% | 49% | 50% |

The estimated number of Radios and TV broadcasts is illustrated in Table 10.2 in four years between 2018 and 2022, which shows the increase in the number of online TVs and blogs to online Radios, Simulcast Radios, and Simulcast TVs. However, the increase does not correlate with the estimated population in the 2022 census. The number of women in the country was estimated to be 31,687,990, which was equal to 51% of the population, while the estimated number of men was 30,053,130, equal to 49% of the total population 61,741,120 (The East African, 2022).

Table 10.2: Number of online radios and TVs

| Year | 2018 | 2019 | 2020 | 2022 |
| --- | --- | --- | --- | --- |
| Online Radio | 7 | 18 | 23 | 25 |
| Online TV | 57 | 216 | 428 | 533 |
| Blogs | 38 | 78 | 118 | 133 |
| Simulcast Radio | 9 | 19 | 25 | 29 |
| Simulcast TV | 1 | 4 | 6 | 9 |

The total subscriptions to the mobile network were 55,293,345, while the subscriptions to the fixed network were 71,894 in March 2022, making the total 55,365,239 of the total population. This implies that approximately 6,375,881, equal to 10% of the total population, do not subscribe to mobile networks. However, the level of preparedness of Tanzania as a country as measures to overcome the effects that could be the outcome of the COVID-19 outbreak and lockdown, that is, the policies and guidelines put in place included: (a) the authorities stopped reporting new cases since early May 2020 to reduce the anxiety among the citizens. (b) Large gatherings were forbidden (except for worship), schools and educational institutions suspended attendance, international flights were cancelled, and wearing of face masks was mandated to reduce the transmission rate (Kangwerema et al., 2021; Mboera et al., 2020; Mugabe et al., 2022). Besides, the lockdown's feasibility, appropriateness, and effectiveness as a mitigation measure varied from place based on the level of the economy (Kangwerema et al., 2021). Mboera et al. (2020) pointed out that the lockdown restrictions varied from country to country based on the severity of the pandemic, such as "loose lockdown" or "tight lockdown" and "total lockdown" or "partial lockdown."

Like many other nations, Tanzania encountered enormous difficulties maintaining educational continuity due to school closures and travel restrictions during the COVID-19 pandemic. Makoye (2020) highlighted specific educational policies and programs that fall under this category to promote online learning and close

Tanzania's digital divide as follows: (1) Tanzania Institute of Education (TIE) released many digital learning aids to assist learners throughout the pandemic. These comprised digital materials and e-books for elementary and secondary schooling that could be accessed via their official website. TIE also sought to improve its digital repository with interactive learning resources to support educators and learners in distance learning settings. (2) The Ministry of Education, Science, and Technology (MoEST) partnered with nearby radio and television stations to broadcast educational programming. Learners in primary and secondary schools were the target audience for these broadcasts, which addressed important topics. Through this program, learners in underserved and remote areas with little to no internet access can now be reached. (3) Shule Direct, an EdTech project initiative, offered digital educational materials that were in line with Tanzania's national curriculum. Through their mobile applications and web platform, they provided resources for teachers and students, such as interactive courses, practice questions, and study guides.

Shule Direct increased its efforts to improve the accessibility of learning resources during the epidemic, especially mobile-based solutions that could be utilized offline. (4) A social enterprise based in Tanzania called Ubongo used its well-liked educational cartoon series, Ubongo Kids, to provide educational materials. Thanks to the series airing on the radio and television, children were guaranteed to continue learning through engaging and instructive content. Ubongo also created mobile applications to reach a varied audience and offered instructional resources in other languages. (5) During the epidemic, MwalimuPlus App provided an online platform with interactive information, quizzes, and video courses for secondary school pupils, making it another noteworthy initiative. With the help of digital resources, the app attempted to offer a whole e-learning experience in addition to traditional classroom instruction. (6) The Tanzanian government modified the Education Sector Development Plan (ESDP) to account for the transition to digital education. This involved prioritizing digital infrastructure and online pedagogy-focused teacher training programs. To evaluate the success of projects for remote learning and make data-driven decisions, efforts were also undertaken to improve the monitoring and evaluation framework.

### STEM teaching system and its change

STEM teaching system, like any other teaching system, is considered to be made of three elements: Knowledge (considered as node "K"), Beliefs (considered as node "B"), and Practice (considered as node "P"). The interaction of the three elements can result in a change in the STEM teaching system. According to Brunetto et al. (2022), STEM teaching practices can be stable and resistant to change as STEM teaching beliefs are. However, several examples are given in the literature, and through the subsisted experiences of various STEM researchers who engage themselves in different professional learning courses, change is possible and can occur quickly if it is enhanced. Kleickmann et al. (2013) pointed out that a change in practice can sometimes lead to a change in beliefs and vice versa. They also reported that when changes in STEM teachers' practice occur, changes to student learning can subsequently be observed, and the allied beliefs about STEM teaching and learning change. This proposes that knowledge, practice, and beliefs be systemically interconnected as a change in one component also ignites a change in other elements (Brunetto et al., 2022). For instance, in STEM teachers' professional learning, when a new topic is studied, or a new experience is shared, it may aggravate the acquisition of knowledge, whereby new facts are acquired and become part of a STEM teacher's repertoire. In the end, changes in teachers' beliefs and practices may occur.

Brunetto et al. (2022) proposed a model that reciprocates the change in the three elements, as seen in Figure 10.1(a). They considered "K", "P", and "B" as the nodes in a graph connected by undirected edges between them to signify their connection and the mutual influence on one another. From Figure 10.1(c), the new beliefs "B¹" can be connected to either new practice "P¹" or old "P" and to either old "K" or new knowledge "K¹". It depends on how the STEM teachers described and lived the change in "K", "B", or "P" (Cutri et al., 2020). Literature has described "B¹" as mostly connected to P¹ because the change in practice (P¹) subsequently provokes a change in beliefs (B¹). Another way to produce changes in belief structures can be when pre-service teachers' experiences have an immediate and profound transformative impact on their beliefs (Brunetto et al., 2022; Cutri et al., 2020). When teachers' knowledge about STEM is changed, a change in their beliefs might be provoked (Kleickmann et al., 2013); thus, the two elements, i.e., "K" and "B", are mutually connected. The mutual change in the two elements can result in new elements such as "K¹" and "B¹". Thus, the interconnection and subsequent change between the three aspects of STEM education can also be used to elaborate on the changes that may occur in teacher educators' and student teachers' knowledge, practices, and beliefs.

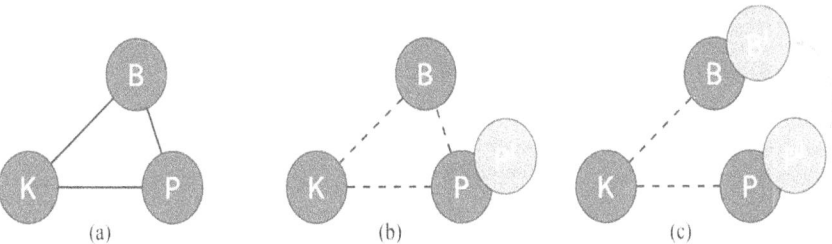

**Figure 10.1:** STEM education and its change.

The mutual change in educators' and learners' knowledge, attitudes, and practice can sometimes be induced by the information they acquire from their professional learning. For instance, Cutri et al. (2020) proposed a prominent way to cause change among STEM teachers during their professional knowledge by involving them as learners in STEM and pedagogy. This can also be applicable in the current context of STEM education after the COVID-19 pandemic (Brunetto et al., 2022). The pandemic resulted in a change in teachers' practice through blended teaching and learning. Besides, Cutri et al. (2020) highlighted that some teachers were more comfortable teaching online than physically, based on the situation that prevailed during the pandemic season. This implies that the new practice impacted these teachers' change in beliefs.

## New visions of STEM teaching after COVID-19
### Resilience building in STEM education

The experiences acquired from the hit of COVID-19 pandemic reveal that building a resilient system in STEM teaching and learning is paramount. Most learners consider STEM subjects difficult and abstract (Xenofontos and Mouroutsou, 2022). People mostly believe they are not good at STEM (Johnston-Wilder et al., 2016), but learners must overcome such negative attitudes toward STEM learning. Such kind of mentality, if not rectified, may worsen, especially when the whole process of STEM learning is being interfered with by disasters like the COVID-19 pandemic (Johnston-Wilder et al., 2016; Xenofontos and Mouroutsou, 2022). Ariyanto et al. (2017) argued that resilient educators have a growth mindset for learning. They are comfortable with challenges and the idea of STEM struggle. However, building resilience by the teachers takes time, but it is achievable in the classroom environment. Johnston-Wilder et al. (2016) suggested that one way to make a resilient learning system is to create an environment that builds a positive learning culture among learners. Moreover, teacher-designed culturally and socially meaningful tasks, instructional practices that support respect for learners' social interactions, and the development of key mathematical practices such as justifying, explaining, and questioning (Ariyanto et al., 2017).

### Encourage learner-centered pedagogy

Pedagogical approaches that engage learners in the learning process and require them to think and participate in meaningful learning activities are considered to be learner-centered. In Partanen's (2018) view, learner-centered approaches shift the focus from the teacher to the learner by accommodating the balance of power in the classroom and the roles of both the teacher and the learners. In addition, re-evaluating learners' responsibilities for learning, reconsidering the function of content to be learned, and the purpose and processes of evaluation (Don et al., 2023; Godber and Atkins, 2021). In STEM teacher education, learner-centered approaches may create a generation of resilient teachers who can confidently handle classes in all learning environments. Therefore, in this perspective, teacher educators are encouraged to build student teachers' capacities to face the challenging teaching environment as individual learners.

## A blend of digital and face-to-face pedagogy

Online learning and digital pedagogy were immediate solutions appropriate to respond to the COVID-19 lockdown in most countries' education systems. The shift was abrupt in that the preparedness of educators and learners was still low (Foster et al., 2022; Murgatrotd, 2020; Rubel et al., 2021). The lockdown in the context of COVID-19 was deemed necessary to protect people from being infected, although it interfered with other aspects of life (DeCoito and Estaiteyeh, 2022; Godber and Atkins, 2021). However, the blend of digital and face-to-face pedagogy is inevitable in sustaining STEM learning in this era, regardless of the circumstances. Ruggiero and Mong (2015) highlighted that effective integration technology in education needs reviewing the current practices and recommending effective strategies for restructuring policies. In this regard, the proposed good practice of technology integration to inform policy should incorporate crisis-ready digital pedagogies.

## Theoretical framework

Crises come in many forms and shapes, which may be difficult to predict or plan for in advance. According to Aspriadis (2021), a crisis is an untimely event that includes the element of surprise and generally marks a phase of disorders in the seemingly normal development of a system. A crisis creates high threats and uncertainty to the organization's high-priority goals (Walby, 2022). Even with the necessity of using and transferring information and technology, it is widespread for the end users of technology in education, that is, educators and learners, to resist such technologies. This study, therefore, was underpinned by the "Crisis Management Theory" proposed by John Burnett in 1998 (Aspriadis, 2021; Walby, 2022), complemented by the Unified Theory of Acceptance and Use of Technology (UTAUT) proposed by Venkatesh et al. (2003). Crisis management is an organization's approach to handling emerging trouble, which is paramount to enable the organization to plan. Besides, Chan et al. (2021) highlighted that crises include contention, risk, accidents, emergencies, and all other uncontrollable problems. Therefore, deliberate efforts should be put in place to combat the crises mentioned above in the context of STEM education and raise the preparedness of bodies in the teacher education system.

The UTAUT model was adopted in this study to establish how STEM teacher educators and student teachers can accept and effectively use (voluntary and involuntary) technology in teaching and learning. Ayaz and Yanartaş (2020) stated that it is widespread for the end users of technology, such as educators and learners, to resist the use of technology. UTAUT comprises four main factors. These are performance expectancy (PE), social influence (SI), effort expectancy (EE), and facilitating conditions (FC), as seen in Figure 10.2. Moreover, it includes four intermediate individual variation variables such as gender, age, experience, and voluntariness of use. The four variables predict the relationship between primary factors, behavioral intention, and user behavior (Ayaz and Yanartaş, 2020). The primary factors, sometimes called determining factors, include PE, SI, EE, and FC (Venkatesh et al., 2003). The FC is empirically identified as the direct determinant of adopting the behavior (Ayaz and Yanartaş, 2020; Venkatesh et al., 2003). Therefore, the above factors are prominent as direct user behavior and acceptance determinants.

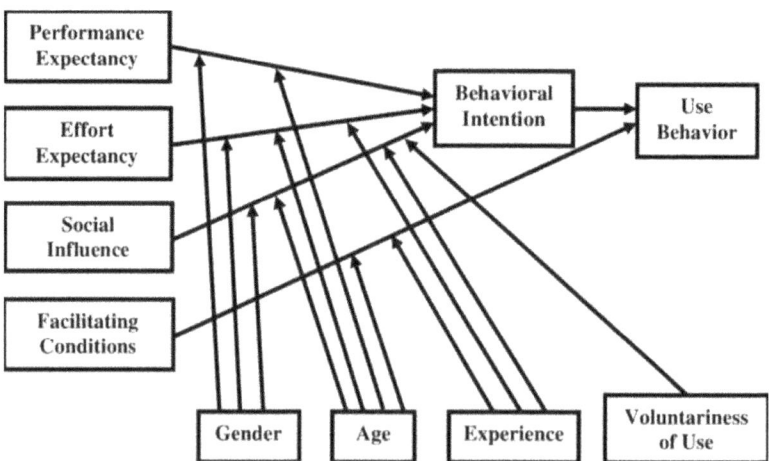

**Figure 10.2:** Unified theory of acceptance and use of technology (UTAUT) model.

The COVID-19 pandemic has paved a comprehensive picture of how teaching and learning systems are from being crisis-ready. The lockdown worldwide left learning institutions, educators, learners, and parents in a puzzle without knowing how teaching and learning could be facilitated. This was a challenge for developing countries like Tanzania, which still need to catch up with the fast-growing world of technology and digital literacy. Even with the challenges above, the government facilitated teacher education colleges with ICT facilities through various interventions and projects, although more is needed to suffice the magnitude of the need. This research was conducted to descriptively interpret the teacher educators' experiences and practices that can enhance the successful existence of STEM education in the new normal after the COVID-19 pandemic. Therefore, this research answered the following research questions:

1. What are the teacher educators' experiences that can be drawn from the outbreak of the COVID-19 pandemic?
2. How can STEM education be contextualized in the outcomes of the new normal after the COVID-19 Pandemic?

## Methodology

This work is a product of the Foundations for Learning (F4L) interventions under the Foundations for Education and Empowerment (F4EE) project, funded by Global Affairs Canada (GAC), and the Aga Khan Foundation, Canada. This research was conducted between January and March 2020 as the Reconnaissance part of the foundation for learning project conducted by the Aga Khan Development Network (AKDN), that is, Aga Khan University—Institute of Education Development (AKU-IED) and Aga Khan Foundation. The study was conducted in the four Districts of Lindi Region, southern Tanzania.

### Research design

This research was designed based on the interpretive worldview (Hothersall, 2019) framed in the qualitative research approach (Creswell and Poth, 2016). The study followed the interpretivism paradigm due to the subjective nature of controversial issues like the outbreak of the COVID-19 pandemic (Hardy, 2016; Hothersall, 2019). According to (Hothersall, 2019), reality is based on the naturalistic understanding of individuals' beliefs, opinions, and social emotions to reason and decode the information that is the outcome of a phenomenon. Besides, the qualitative nature of the study enabled the best interpretation of the teacher educators' experiences with COVID-19 and drew meaning that cannot be obtained through measurement of the variables (Creswell and Poth, 2016; Reid et al., 2014).

### Research methods and procedures

In this study, data were collected using a semi-structured Focus Group Discussion (FGD) guide, whereby necessary ethical issues were considered based on the AKU Ethical Review Committee (ERC) before the commencement of data collection procedures. The study involved 22 teacher educators who were purposively selected to participate in this study. The selected teacher educators participated in three FGDs at their convenience. Two groups were made of seven, and one had eight teacher educators.

## Data analysis procedures

The FGD instrument employed in this research was checked for credibility by experienced qualitative researchers in STEM and teacher education to establish inter-rater reliability and ascertain its worthiness to yield credible research outcomes. The inter-rater reliability was established by the external qualitative researchers "raters" who rated the instruments and recommended some changes. The recommendations made by the raters were discussed by the research team and were effected before data collection after reaching a consensus. Moreover, member checking was done to observe the conformability of the information obtained from the participants (Nowell et al., 2017). Audit trials were also performed during data analysis to safeguard an accurate portrayal of the study findings using the respondents' views (Creswell, 2014).

Furthermore, data analysis in this study was done daily as the data was collected to reduce the bulkiness of the data collected (Creswell, 2014). Constant reflection on FGD information was done to closely check the ongoing data collection process and identify issues that needed follow-up to inquire for clarity from the participants. Meanwhile, data analysis was performed thematically (Braun et al., 2016), beginning with data analysis (Braun et al., 2016) and beginning with the transcription of the audio data and field notes. Then, the transcribed data was organized based on their types to form a database for the inductive coding process (Nowell et al., 2017). Then, themes were generated from the coded data, sorted, and sifted through to identify similar and coherent phrases aligned with the research question (Braun et al., 2016). Finally, interpretation was made to generate meaningful information to answer the proposed research questions.

## Research Findings

The results from three FGDs conducted with pre-primary and primary education teacher educators were organized to generate excerpts reflecting their experiences from the COVID-19 pandemic and foresee the contextualization of these experiences to enhance the teacher education system coping with the new normal.

### Experiences drawn from the COVID-19 pandemic

Teacher educators were asked about their perceptions of the interferences that resulted from the COVID-19 pandemic. They highlighted that the lockdown was inevitable, but the only problem was its interference with the planned learning outcomes because they were no longer supposed to contact student teachers for face-to-face teaching and learning.

> ... the response to the lockdown by the teacher educators and student teachers was positive because they understood the magnitude of the problem. However, the problem was only the interference in the timetable and the content anticipated to be learned because they had already planned from the beginning of the year. Nevertheless, the CORONA virus interrupted everything. (TE 3: FGD3)

After the reopening of the college, safety measures were highly encouraged to be observed by both student teachers and their teacher educators during learning to avoid spreading the virus. Teacher educators were advised to use strategies that do not make many students and teachers come in contact simultaneously. One of them said:

> ... as what the previous speaker has said, interactions between the student teacher with another student teacher decreased because of the social distancing that is needed bearing in mind the rules about prevention against the spread of the virus (TE 5: FGD 1)

The lockdown and its interference with STEM teaching and learning resulted in student teachers' low performance in various classroom activities and some formative examinations. This was due to the three months that most of them did not access learning; even for those with access to technology, learning was not continuous. In addition, the student teachers were still anxious about the spread of the virus; as a result, their motivation to learn also went low.

> Their performance went down because, at the time they were supposed to be in college, they were not there because of COVID-19. So, when they came back, that speed and motivation for studying declined to a large extent, which is why the performance also declined due to COVID-19. (TE 3: FGD 2)

Similarly, the STEM teacher educators highlighted that sharing the available resources while observing social distancing was challenging for student teachers, considering the ratio between their number and the available resources. There was a restriction on the number of student teachers who were supposed to be simultaneously in the library and the computer lab. For instance, it was mentioned that the computer lab only has thirty-two computers. The student teachers enrolled for pre-primary and primary courses are more than 50 each year. With this number, it was difficult for the whole class to use the computer lab simultaneously. One of the teacher educators during FGDs said:

> … if the library can accommodate 50 people, this means that during COVID-19, we were forced to reduce the number to observe social distancing. It is like if you were letting 50 people, that means you need to allow 25 people. If you have around 300 student teachers, you can see how big the circle is for everyone to get the right to use the library. (TE 2: FGD 3)

## Preparedness to blend face-to-face and digital pedagogy

Based on the impact that COVID-19 had on STEM teaching and learning and the teacher educators' experiences, a blend of face-to-face and digital pedagogy was considered a viable measure in response to a similar crisis in the future. This came out during the discussions with teacher educators, although the discussions revealed that student teachers and teacher educators still needed professional help in effectively using digital pedagogy. Some teacher educators viewed preparedness as an initiative that the government and the college administration took. The initiatives could include workshops, training, and various interventions that may enrich teacher educators' and student teachers' capacities in integrating digital pedagogy in STEM teaching and learning.

> How prepared is the college regarding risks of closure due to crisis like COVID-19 … for instance, plans to provide online teaching and learning in case of any physical restrictions – knowledgebase, materials, and infrastructure? (Researcher)

One of the teacher educators replied:

> The college is not prepared yet. It is not prepared because if they were prepared, we could have a different program that the ICT experts have designed for teaching. (TE 1: FGD 1)

Another teacher educator in the same FGD complemented:

> Maybe to clarify why our colleague has said the college is not prepared, I can say that although we are provided with ICT facilities, it is still difficult to effectively implement online learning because it requires continuous interaction between the teacher educators and student teachers who come from different places where the network is a problem, … also it is not all who have the gadgets such as laptops, smartphones, and internet access, especially student teachers. Therefore, it will still be difficult for the college even if the college has a plan. It could still have been a problem for student teachers regarding effective lesson delivery. (TE 6: FGD 1)

Apart from the initiatives the higher authorities can take, teacher educators revealed their readiness to advance teaching and learning STEM using the available technological resources.

> I think some of the teacher educators individually initiated some programs through which they were sharing materials with students, asking questions, giving answers to students' posed questions, etc.… through some online system like WhatsApp, but still more training is required to enrich their understanding and improve their expertise in the use of varieties of online learning systems. (TE 4: FGD 3)

The concern was on the student teachers who come from different economic backgrounds. It was mentioned that some student teachers could not afford to own personal gadgets like tablets and smartphones. Moreover, internet access when out of the college premises could be a challenge to some of them due to financial constraints.

> But let us say using WhatsApp, but you will be teaching a few individuals how many have smartphones that they can use to access WhatsApp or Facebook and be able to access other programs by using the smart phones in this case, … even if you use WhatsApp still big group will be left out to participate in the program. (TE 6: FGD 2)

> … perhaps this needs strong measures to enable us to succeed. If possible, the Ministry or various donor agencies can intervene to ensure that each student teacher has gadgets like a tablet or a smartphone to enhance their online learning when they report at the college … these can help easy interaction and communication even when they are out of college premises. (TE 2: FGD 2)

> … for us to be prepared for facing/confronting/dealing with crises like COVID-19 is not difficult, as I said in my first response. We can be prepared, but what about our clients … the student teachers? I mean the facilities … do they have facilities to help them to access those lessons? Is there a network connection between them and the teacher educators? The computer lab has internet access for our college, but sometimes it is overloaded, and the network becomes weak to be shared by all students. (TE 4: FGD 1)

> … also, the use of ICT sometimes, instead of giving them activities in groups, activities are given through WhatsApp groups instead of normal groups where people sit together and discuss they use different programs in the internet. (TE 8: FGD 3)

## Use of active learner-centered learning

It was suggested during the discussions that student-teacher still need to be capacitated to take charge of their learning apart from regular face-to-face learning with digital pedagogy. The teacher educators proposed various active learner-centered teaching and learning strategies to supplement online learning. They said that:

> Maybe another strategy that can help the student teacher deal with the challenges is using various active learning strategies. They may outsource materials from various sources to prepare for any activity given. When learning, I can propose the use of any strategy, for instance, the use of gallery walk … this is a strategy that makes the student-teacher first participate in the activity which s/he has been given because s/he will be asked different questions which will require them to explain. (TE 7: FGD 1)

In addition, two other teacher educators concurred and said:

> … by doing so, it helps him/her to solve different learning challenges which s/he has, and others also will contribute, so it helps him/her to understand better if there is an area which s/he did not understand well and others will get knowledge on the learning content that has been prepared. (TE 2: FGD 1)

> *Strategies to develop collaborative and interactive teaching and learning should be used. The teacher can prepare activities that involve student-teachers more in preparing and implementing the lesson's content. It may require them to do and talk so they can also be seen participating in something taught then. Therefore, during preparation as a teacher, you must prepare activities that require them to be more performers than only being listeners, and you as a teacher should be a facilitator to make sure they do not go out of the intended learning outcome.* (TE 8: FGD 2)

## Discussion

In light of the first research question, the findings indicate that the lockdown, which was the outcome of the COVID-19 pandemic outbreak, gave multiple experiences to STEM teacher educators. Because formal face-to-face classroom sessions were seized during this season, teaching and learning should continue from home. Various ways were used to make learning continuous, especially using technological devices. One of the prominent ways the teacher educators in the study reached their student teachers during the pandemic was through WhatsApp. They used WhatsApp to interact in groups, send them activities to be done at home, and later submit for assessment. This finding brings food for thought that online learning or online interaction between STEM teacher educators and student teachers can go beyond WhatsApp as a platform (Drijvers et al., 2021; Jaramillo-Arévalo et al., 2023). In line with this finding, Drijvers et al. (2021) highlighted in their research that teachers tend to use standard digital tools that existed even before COVID-19 for the delivery of lessons and communication with their learners. In addition, Video usage emerged as a component of technology use that substantially impacts learners' learning experiences in STEM (Attard and Holmes, 2022). To complement this finding, Rubel et al. (2021) reported videos to be more stimulating than traditional writings, and they prized the ability to re-watch the videos sourced from YouTube, which added a different outlook on the STEM content. They also proposed other forms of digital technology that may facilitate STEM lessons, such as video conferencing and other free online digital resources. With the mentioned findings, therefore, it is evident that there are varieties of digital technology in which STEM teaching and learning can be successfully and effectively facilitated.

Moreover, the findings reveal that teacher educators' beliefs directly impacted changing their practices. For instance, teacher educators believe that the only effective digital resource that can enhance learning and student-teacher interaction is the smartphone through WhatsApp. If the teacher is confident with the idea of online pedagogy because he has the right equipment, he strengthens his beliefs and, as a result, changes his practice (Brunetto et al., 2022). Besides, teacher educators should be exposed to various digital resources to change their beliefs, limiting their practice (Chiu and Liu, 2023; Cutri et al., 2020). Further, change in teacher educators' beliefs may positively impact the readiness to use digital technology in teaching and learning. This is similar to Venkatesh et al. (2003), the proponents of the UTAUT framework, which underpinned this research, highlighted that individuals' use of digital technology can be impacted by variables like individuals' experience and voluntariness. Moreover, the findings indicate that teacher educators and student teachers dedicated their efforts to mainstream digital technology in their face-to-face classroom sessions, termed by Ayaz and Yanartaş (2020). Further, the findings highlight that both teacher educators' and student teachers' acceptance and use of varieties of digital resources and technology can be highly strengthened through various technological training, workshops, and interventions; through such kind of initiatives, their knowledge of technology use is strengthened (Brunetto et al., 2022; Cutri et al., 2020).

To answer the second research question, teacher educators' excerpts indicate that adapting digital technology for their student teachers' learning makes it pray its central role while communication becomes key to learning. According to them, student teachers can use digital resources like the internet to search for information related to the learning activities given to them. Additionally, sophisticated mathematics computer systems like Geogebra and STACK allow answers to be evaluated mathematically (Foster et al., 2022; O'Connor et al., 2023). Apart from the use of technology, the findings indicate that shifting the paradigm from teacher-centered to learner-centered can make learning continuous even in the teacher's absence. Shifting the focus to coordinating students' actions and getting them to work individually and then with each other becomes much more important in students' learning (Foster et al., 2022).

Similarly, the findings observed that using various active learning learner-centered techniques can positively impact student teachers' learning. Student teachers can work on STEM activities in small groups or individually and then present to their fellows through strategies like gallery walks. This learning strategy can enable student teachers to learn from their peers' exhibited work (Johnston-Wilder et al., 2016). Thus, active learner-centered STEM teaching and learning techniques can create a resilient generation of prospective pre-primary and primary STEM teachers.

Furthermore, if teacher preparation programs prepare resilient teachers, they can make STEM learning crisis-ready by making learners explore their learning under minimal teacher support (Corpuz, 2021; Foster et al., 2022). However, the highlights from the findings indicate the preparedness of most STEM teacher educators who participated in FGDs to be moderate at the individual level, calling for more efforts from the government and other education stakeholders. This is because enhancing the education system to be crisis-ready is paramount, considering the damages caused by the COVID-19 lockdown (Godber and Atkins, 2021).

## Limitations of the Study

Although this study offers insightful information about teacher educators' experiences with in-person and digital pedagogy in STEM education, some limitations must be noted: First, participants were primarily selected from a particular geographic area, and the study's sample size was constrained. This could restrict how broadly the results can be applied to different people or circumstances. In addition, the majority of the data gathered through FGDs was self-reported, which may create bias. It is possible that participants gave answers that were socially acceptable or that their memories of the events were inaccurate. The study's exclusive focus on the viewpoints of teacher educators may have obscured student experiences and feedback, which are vital for a comprehensive understanding of the relative merits of face-to-face and digital teaching. This variability was not controlled in the study. On the other hand, the study did not extensively explore other contextual factors, such as institutional support, professional development opportunities, or specific curriculum demands, which might impact the implementation and perception of digital and face-to-face teaching methods.

## Future Directions for Research

In light of the limitations noted, additional research in this field could take several approaches to expand on the study's findings: To improve the generalizability of the results, future research should strive to include a more extensive and more varied sample of individuals from different institutional types, educational levels, and geographic regions. Incorporating students' viewpoints in future research will provide a more balanced understanding of the advantages and disadvantages of digital versus in-person pedagogy in STEM education. On the other hand, comparative studies that look at various hybrid teaching models, including varied ratios of digital to in-person education, are another possible research avenue. These studies could offer more profound insights into the most effective combinations of STEM courses and learning outcomes. Examining how focused professional development initiatives affect teacher educators' competence and self-assurance in utilizing digital resources and techniques may also point to practical ways to improve digital pedagogy abilities. Future studies should examine the application and effects of cutting-edge technologies, such as virtual reality (VR) and artificial intelligence (AI), in STEM education as technology advances.

Similarly, researchers should consider how other elements like institutional support, policy frameworks, and resource allocation shape digital and in-person pedagogy. Doing so will shed light on the systemic aspects that affect educational practices. In addition, future research could profit from multidisciplinary approaches that integrate knowledge from the fields of management, technology, psychology, and education to create more comprehensive plans for successful STEM teaching in the new normal. Future studies can help develop a more sophisticated and useful knowledge of the relationships between digital and in-person pedagogy in STEM education by focusing on these areas.

## Conclusion

This study shows that confronting an emergency crisis like COVID-19 in STEM teacher education requires changing knowledge, beliefs, and practices. At this point, it appears that the transformations caused by COVID-19 created the "new normal" that STEM teacher educators and student teachers had to adjust to cope with the teaching and learning in its transformed context. In line with this study's theoretical perspective, the findings suggest that deliberate measures should be implemented to enhance technology acceptance and use by STEM teacher educators and student teachers. The training, workshops, and interventions that may change their beliefs on acceptance and use of digital pedagogy and later change their knowledge and practice are highly advocated for. Besides, mainstreaming digital pedagogy in curriculum and education policies is paramount to enhancing its reflection in teacher and student teachers' practices. Finally, face-to-face and digital pedagogy will be identified as good practices to advance STEM teaching and learning.

## Acknowledgments

We highly appreciate the financial support provided by Global Affairs Canada (GAC) through the Aga Khan Development Network (AKDN) agencies, namely the Aga Khan University—Institute for Educational Development, East Africa (AKU-IED, EA) and Aga Khan Foundation (AKF). We also thank all the F4L research team and the F4L project team for their authoritative support of this project.

## Bibliography

Allen, P., & Trinick, T. (2021). Agency–structure dynamics in an indigenous mathematics education community in times of an existential crisis in education. *Educational Studies in Mathematics 108*(1–2), 351–368. <https://doi.org/10.1007/s10649-021-10098-1>.

Ariyanto, L., Herman, T., Sumarmo, U., & Suryadi, D. (2017). Developing mathematical resilience of prospective math teachers. *Journal of Physics: Conference Series 895*, 012062. <https://doi.org/10.1088/1742-6596/895/1/012062>.

Aspriadis, N. (2021). Managing COVID-19 pandemic crisis: The case of Greece. *Journal of International Crisis and Risk Communication Research 4*(2), 387–412. <https://doi.org/10.30658/jicrcr.4.2.8>.

Attard, C., & Holmes, K. (2022). An exploration of teacher and student perceptions of blended learning in four secondary mathematics classrooms. *Mathematics Education Research Journal 34*(4), 719–740. <https://doi.org/10.1007/s13394-020-00359-2>.

Ayaz, A., & Yanartaş, M. (2020). An analysis of the unified theory of acceptance and use of technology theory (UTAUT): Acceptance of electronic document management system (EDMS). *Computers in Human Behavior Reports 2*, 100032. <https://doi.org/10.1016/j.chbr.2020.100032>.

Braun, V., Clarke, V., & Weate, P. (2016). Using thematic analysis in sport and exercise research. In *Routledge Handbook of Qualitative Research in Sport and Exercise* (pp. 213–227). Routledge.

Brunetto, D., Bernardi, G., Andrà, C., & Liljedahl, P. (2022). Teaching as a system: COVID-19 as a lens into teacher change. *Educational Studies in Mathematics 110*(1), 65–81. <https://doi.org/10.1007/s10649-021-10107-3>.

Chan, M. C. E., Sabena, C., & Wagner, D. (2021). Mathematics education in a time of crisis—A viral pandemic. *Educational Studies in Mathematics 108*(1–2), 1–13. <https://doi.org/10.1007/s10649-021-10113-5>.

Chiu, H.-Y., & Liu, C. (2023). Do STEM teachers have the potential to become leaders in online education? *Journal of Research in STEM Education 9*(1), 1–19. <https://doi.org/10.51355/jstem.2023.123>.

Corpuz, J. C. G. (2021). Adapting to the culture of 'new normal': An emerging response to COVID-19. *Journal of Public Health 43*(2), e344–e345. <https://doi.org/10.1093/pubmed/fdab057>.

Creswell, J. W. (2014). *Research Design: Qualitative, Quantitative, and Mixed Methods Approaches* (4th ed). SAGE Publications.

Creswell, J. W., & Poth, C. N. (2016). *Qualitative Inquiry and Research Design: Choosing Among Five Approaches*. Sage Publications.

Cutri, R., Mena, J., & Whiting, E. (2020). Faculty readiness for online crisis teaching: Transitioning to online teaching during the COVID-19 pandemic. *European Journal of Teacher Education 3*(4), 523–541.

DeCoito, I., & Estaiteyeh, M. (2022). STEM teachers' experiences with online teaching during the COVID-19 pandemic: A Canadian context. In H. Burgsteiner & G. Krammer (Eds.), *Impacts of COVID-19 Pandemic's*

*Distance Learning on Students and Teachers in Schools and in Higher Education—International Perspectives* (pp. 421–444). Leykam Buchverlag. <https://doi.org/10.56560/isbn.978-3-7011-0496-3_20>.

Don Anton Robles Balida et al. (2023). Teaching Pedagogies in the New Normal. *Russian Law Journal 11*(3). <https://doi.org/10.52783/rlj.v11i3.2054>.

Drijvers, P., Thurm, D., Vandervieren, E., Klinger, M., Moons, F., van der Ree, H., Mol, A., Barzel, B., & Doorman, M. (2021). Distance mathematics teaching in Flanders, Germany, and the Netherlands during COVID-19 lockdown. *Educational Studies in Mathematics 108*(1–2), 35–64. <https://doi.org/10.1007/s10649-021-10094-5>.

Foster, C., Burkhardt, H., & Schoenfeld, A. (2022). Crisis-ready educational design: The case of mathematics. *The Curriculum Journal 33*(4), 519–535. <https://doi.org/10.1002/curj.159>.

Godber, K. A., & Atkins, D. R. (2021). COVID-19 impacts on teaching and learning: A collaborative autoethnography by two higher education lecturers. *Frontiers in Education 6*, 647524. <https://doi.org/10.3389/feduc.2021.647524>.

Hamilton, L., Kaufman, J., & Diliberti, M. (2020). *Teaching and Leading Through a Pandemic: Key Findings from the American Educator Panels Spring 2020 COVID-19 Surveys*. RAND Corporation. <https://doi.org/10.7249/RRA168-2>.

Hardy, M. (2016). 'I know what I like, and I like what I know': Epistemology in practice and theory and practice again. *Qualitative Social Work 15*(5–6), 762–778.

Hothersall, S. J. (2019). Epistemology and social work: Enhancing the integration of theory, practice and research through philosophical pragmatism. *European Journal of Social Work 22*(5), 860–870.

Jaramillo-Arévalo, M., Alvarez-Risco, A., De-La-Cruz-Diaz, M., Anderson-Seminario, M. D. L. M., & Del-Aguila-Arcentales, S. (2023). Digital tools to promote STEM education in new normality. In A. Alvarez-Risco, M. A. Rosen, & S. Del-Aguila-Arcentales (Eds.), *Advanced Series in Management* (pp. 107–121). Emerald Publishing Limited. <https://doi.org/10.1108/S1877-636120230000030016>.

Johnston-Wilder, S., Pardoe, S., Almehrz, H., Evans, B., Marsh, J., & Richards, S. (2016). *Developing Teaching for Mathematical Resilience in Further Education*, 3019–3028. <https://doi.org/10.21125/iceri.2016.1652>.

Kangwerema, A., Thomas, H., Knovicks, S., Safari, J., Diluxe, M., Madadi, S., Elhadi, Y. A. M., Ahmadi, A., Adebisi, Y. A., & Lucero-Prisno, D. E. (2021). The challenge of dearth of information in Tanzania's COVID-19 response. *Journal of Global Health Science 3*(2), e20. <https://doi.org/10.35500/jghs.2021.3.e20>.

Kleickmann, T., Richter, D., Kunter, M., Elsner, J., Besser, M., Krauss, S., & Baumert, J. (2013). Pedagogical content knowledge and content knowledge of mathematics teachers: The role of structural differences in teacher education. *Journal of Teacher Education 64*, 90–106.

Kollosche, D., & Meyerhöfer, W. (2021). COVID-19, mathematics education, and the evaluation of expert knowledge. *Educational Studies in Mathematics 108*(1–2), 401–417. <https://doi.org/10.1007/s10649-021-10097-2>.

Makoye, K. (2020). *Tanzania: Lockdown Brings Digital Innovations in Learning COVID-19 Lockdown Expedites Innovative Technology, but also Spotlights Digital Divide in East Africa.* <https://www.aa.com.tr/en/africa/tanzania-lockdown-brings-digital-innovations-in-learning/1933192>.

Mboera, L. E. G., Akipede, G. O., Banerjee, A., Cuevas, L. E., Czypionka, T., Khan, M., Kock, R., McCoy, D., Mmbaga, B. T., Misinzo, G., Shayo, E. H., Sheel, M., Sindato, C., & Urassa, M. (2020). Mitigating lockdown challenges in response to COVID-19 in Sub-Saharan Africa. *International Journal of Infectious Diseases 96*, 308–310. <https://doi.org/10.1016/j.ijid.2020.05.018>.

Mugabe, P. A., Renkamp, T. M., Rybak, C., Mbwana, H., Gordon, C., Sieber, S., & Löhr, K. (2022). Governing COVID-19: Analyzing the effects of policy responses on food systems in Tanzania. *Agriculture & Food Security 11*(1), 47. <https://doi.org/10.1186/s40066-022-00383-4>.

Murgatrotd, S. (2020). *COVID-19 and Online Learning.* <https://doi.org/10.13140/RG.2.2.31132.85120>.

Nowell, L. S., Norris, J. M., White, D. E., & Moules, N. J. (2017). Thematic analysis: Striving to meet the trustworthiness criteria. *International Journal of Qualitative Methods 16*(1), 160940691773384. <https://doi.org/10.1177/1609406917733847>.

O'Connor, J., Ludgate, S., Le, Q.-V., Le, H. T., & Huynh, P. D. P. (2023). Lessons from the pandemic: Teacher educators' use of digital technologies and pedagogies in Vietnam before, during and after the COVID-19 lockdown. *International Journal of Educational Development 103*, 102942. <https://doi.org/10.1016/j.ijedudev.2023.102942>.

Partanen, L. (2018). Student-centred active learning approaches to teaching quantum chemistry and spectroscopy: Quantitative results from a two-year action research study. *Chemistry Education Research and Practice 19*(3), 885–904. <https://doi.org/10.1039/C8RP00074C>.

Pokhrel, S., & Chhetri, R. (2021). A literature review on impact of COVID-19 pandemic on teaching and learning. *Higher Education for the Future 8*(1), 133–141. <https://doi.org/10.1177/2347631120983481>.

Reid, A. D., Hart, E. P., & Peters, M. A. (Eds.). (2014). *A Companion to Research in Education*. Netherlands: Springer. <https://doi.org/10.1007/978-94-007-6809-3>.

Rubel, L. H., Nicol, C., & Chronaki, A. (2021). A critical mathematics perspective on reading data visualizations: Reimagining through reformatting, reframing, and renarrating. *Educational Studies in Mathematics 108*(1–2), 249–268. <https://doi.org/10.1007/s10649-021-10087-4>.

Ruggiero, D., & J. Mong, C. (2015). The teacher technology integration experience: Practice and reflection in the classroom. *Journal of Information Technology Education: Research 14*, 161–178. <https://doi.org/10.28945/2227>.

Sullivan, P., Bobis, J., Downton, A., Feng, M., Hughes, S., Livy, S., McCormick, M., & Russo, J. (2020). Threats and opportunities in remote learning of mathematics: Implication for the return to the classroom. *Mathematics Education Research Journal 32*(3), 551–559. <https://doi.org/10.1007/s13394-020-00339-6>.

TCRA. (2022). *Communications Statistics Quarter 3 -2021/2022 March 2022*. Tanzania Communications Regulatory Authority. <https://www.tcra.go.tz/uploads/text-editor/files/QUARTERLY%20COMMUNICATIONS%20STATISTICS%20-%20March%202022%20Report_1653820853.pdf>.

The East African. (2022). *Tanzania population (2022 census results)*. <https://www.theeastafrican.co.ke/tea/news/east-africa/tanzania-population-expanded-16-million-in-decade-4003958#:~:text=Monday%20October%2031%202022&text=President%20Suluhu%20said%20the%20census,and%2030%20million%20are%20males>.

Venkatesh, M., Davis, & Davis. (2003). User acceptance of information technology: Toward a unified view. *MIS Quarterly 27*(3), 425. <https://doi.org/10.2307/30036540>.

Videla, R., Rossel, S., Muñoz, C., & Aguayo, C. (2022). Online mathematics education during the COVID-19 pandemic: Didactic strategies, educational resources, and educational contexts. *Education Sciences 12*(7), 492. <https://doi.org/10.3390/educsci12070492>.

Walby, S. (2022). Crisis and society: Developing the theory of crisis in the context of COVID-19. *Global Discourse*, 1–19. <https://doi.org/10.1332/204378921X16348228772103>.

Xenofontos, C., & Mouroutsou, S. (2022). Resilience in mathematics education research: A systematic review of empirical studies. *Scandinavian Journal of Educational Research*, 1–15. <https://doi.org/10.1080/00313831.2022.2115132>.

CHAPTER 11

# Mathematics Teaching During and After Times of Crisis in the African Higher Education Context

*Shonisani Agnes Mulovhedzi,*[1] *Nosisi Nellie Feza,*[1] *and Tawanda Runhare*[1]
[1]University of Venda

**ABSTRACT**
This chapter explored how Foundation Phase lecturers and student teachers coped with the Teaching, learning, and assessment of Mathematics during and after the COVID-19 pandemic, which forced a lockdown of all socio-economic activities, including all education sectors. This chapter discussed the lecturers' typical teaching approaches that took place online. The authors used qualitative methods to get in-depth information to explore mathematics teaching, learning, and assessment during the crisis. The study was underpinned by Venkatesh, Thong, and Xu's 8-point model on the Extended Unified Theory of Acceptance and Use of Technology, which predicts behavioral acceptance of technological innovations. Ten Foundation Phase Mathematics student teachers and three lecturers who taught Mathematics modules using blended pedagogies were interviewed. The lesson observations of the three mathematics lecturers were conducted. The findings revealed the challenges lecturers and student teachers encountered while teaching and learning mathematics, such as inadequate resources, a lack of commitment, and unreliable internet services. The chapter highlights how lecturers can design their Mathematics class activities using blended pedagogies for optimum teaching, learning, and assessment in the Foundation Phase. It concludes with the importance of blended pedagogies in tandem with the fourth industrial revolution.

*Keywords:* Blended teaching and learning, COVID-19 pandemic, digitalization, mathematics education, online teaching

## Introduction and Background

The COVID-19 pandemic has had a profound impact on higher education worldwide. As universities moved to remote learning, many students and instructors needed help to adapt to this new mode of education. Globally, higher education institutions were forced to close their campuses because of the COVID-19 pandemic. The pandemic also interrupted teaching, learning, and assessment in various institutions during and after COVID-19. The Department of Higher Education and Training introduced online learning to be used by lecturers and students. Krishnamurthy (2020) concurs that the effect of online teaching in higher education institutions is significant and monumental. The COVID-19 pandemic altered the lives of students in a variety of ways, including displacing them from their homes and campuses, causing them to face financial difficulties, preventing them from participating in internships, and requiring them to acquire knowledge of new technologies in addition to the subject matter of their studies (Govindarajan and Srivastava, 2020). Mathematics, in particular, is a subject that requires a high level of interaction between lecturers and students. Therefore, the shift to remote learning has particularly challenged Mathematics education. During times of crisis, such as the COVID-19 pandemic, universities must adapt their Mathematics teaching strategies to ensure students can continue learning effectively. One way to do this is through online platforms and technology. The lecturers can use video conferencing software to conduct live online classes and adopt digital whiteboards and other tools to help students visualize and understand mathematical concepts.

In December 2019, Wuhan, which is in China, was the first city to record a case of the COVID-19 viral epidemic. On January 10, 2020, the World Health Organization (WHO) declared the outbreak a Public Health Emergency of International Concern. On March 11, 2020, the WHO declared the epidemic a global pandemic. In reaction to this news, educational facilities worldwide were forced to close, accelerating the demand for education provided via remote means. As a result of the COVID-19 pandemic, over 1.5 billion pupils, equivalent to 90 % of all enrolled students, could not attend classes. South Africa reported its first case of COVID-19 on March 5, 2020; on March 27, 2020, the entire country went into lockdown to stop the virus from spreading further. The COVID-19 pandemic had a significant impact on education worldwide, including the teaching of Mathematics. With the closure of schools and universities in many countries, lecturers were forced to transition to online teaching and remote learning, which posed various challenges for teaching Mathematics.

The pandemic also interrupted teaching, learning, and assessment in various institutions during and after COVID-19. Numerous educational establishments were compelled to find and implement various strategies that helped sustain their academic projects. These strategies included, but were not limited to, emergency remote learning and teaching, making working-from-home arrangements for staff, discovering alternative ways to support students, and reallocating budgets to address emerging needs. All of these were necessary steps in the process.

## Literature Review

Mathematics education in higher education is frequently shaped by many elements, from technical progress to socio-economic changes. Conversely, crises present unique opportunities and problems for lecturers and students. Examples of these include natural disasters, pandemics, and political instability. However, they can also stimulate creativity and resilience among institutions and individuals. This literature review investigates the influence of different crises on mathematics teaching at universities, analyzing the immediate reactions and long-term approaches implemented to preserve the continuity and efficacy of education. This review aims to interpret research findings to emphasize the adaptive strategies, technological interventions, and pedagogical innovations developed in response to these crises. It provides valuable insights into the future of mathematics education in the higher education system.

## Impact of COVID-19 lockdown on the education sector

The COVID-19 epidemic has significantly affected the education industry worldwide, including the higher education system in Africa. The pandemic-induced lockdowns and limitations caused significant disruptions to conventional teaching and learning methods, especially in modules like mathematics that heavily depend on face-to-face interaction and hands-on involvement. The diverse effects of the COVID-19 lockdown on mathematics education in African institutions, both during the crisis and in the period that followed. The COVID-19 epidemic has had a significant impact on higher education. As a result, educational institutions and lecturers must adapt to new learning and teaching settings online. The use of forced remote teaching and learning, which Dhawan (2020) sees as preparing the groundwork for the implementation of digital learning, exemplifies a paradigm shift in how lecturers provide students with a quality education.

## Socio-economic influence on digital learning

Current literature indicates economic, social, and psychological repercussions on students' lives while they are away from their regular schedule of study (Pokhrel and Chhetri, 2021). On-campus activities were prohibited at educational institutions, and lecturers were instructed to make blended and online learning opportunities available to students. They were also instructed to implement track-and-trace options, social distancing, and high levels of hygiene (Godber and Atkins, 2021). In some higher education universities, these limits mandate that all lectures be presented digitally, primarily through the university's online learning management system (Blackboard or Moodle). This unanticipated transition to remote teaching was made difficult because emergency remote teaching in response to a crisis shows minimal resemblance to online teaching and learning that has been purposefully prepared (Scherman, 2020).

## Mathematical models for online learning

The shift from traditional classroom-based Mathematics teaching to online instruction brought several issues. As a highly participatory and problem-solving field, Mathematics has traditionally relied on face-to-face interactions, visual aids, and physical manipulatives to teach its students. The absence of these components in remote teaching necessitates the development of novel approaches to the problem. For lecturers to engage students successfully, encourage active learning, and facilitate experiences in collaborative problem-solving, they needed to create new ways to do it in virtual worlds. In addition, synchronous and asynchronous learning strategies were blended to maximize flexibility and sustain active participation from students. Live virtual sessions were supplemented by pre-recorded lectures, video tutorials, and online resources, allowing students to review subjects at their own pace and in their own time (Blankman, 2020).

## Interventions to institutional closures due to COVID-19

In these times of the COVID-19 pandemic, online curriculum delivery has become a crucial teaching style and a prerequisite to enable continuous teaching and learning across several universities in South Africa. Up to March 2020, the typical educational setting at most South African institutions consisted of students gathering in lecture halls and lecturers employing conventional curriculum delivery. This typical teaching setting is crucial in teacher training programs because it allows lecturers to illustrate and model the vital abilities that student teachers should learn and possess for their profession. Because of the national lockdown and the subsequent cancellation of university classes requiring students' physical presence, students and lecturers were forced to deal with an entirely new circumstance, which, for many educational institutions, represented an untested method of instruction (Huber and Helm, 2020; Eickelmann and Gerick, 2020). Even though the transition to online curriculum delivery made it possible to continue teaching and learning, thanks to the utilization of a variety of digital tools and resources, the most important concern is the degree to which the alternative means of schooling affect the quality of curriculum delivery, the conceptual development of students, and the development of skills that are essential to specific professions.

## Mode of curriculum delivery during the COVID-19 pandemic

The distance learning approach to teaching mathematics is most pertinent to the COVID-19 epidemic. The provision of infrastructure to support lecturers, institutions, and students makes its implementation difficult. There are two categories for online learning recipients: groups of students who use semi-online learning, such as giving assignments through a WhatsApp group without face-to-face, and groups of students who cannot be forced to use online learning due to a lack of infrastructure and technological support, such as in remote locations with no electricity, spotty signal or no facilities for gadgets or laptops.

According to earlier research on online curriculum delivery, there may be new educational opportunities provided by digital technologies (Li and Ma, 2010; Chauhan, 2017), and higher education institutions have increasingly embraced Information and Communication Technology (ICT) integration in recent years. They further observe that the existence of computer equipment for lecturers and students may not ensure students' understanding or advancement despite the potential impact that using ICT tools may have on learning and teaching (Li and Ma, 2010). We believe that far-reaching added value regarding digital literacy competencies among instructors and pupils may not yet be assured in South Africa, which is marked by racial, social class, and habitat disparities.

Since the COVID-19 pandemic, arguably, there has been adequate research on lecturers' experiences of online curriculum delivery during and after the crisis, which focused on mathematics education. This demonstrates the value of lecturers' assessment of the effectiveness of teaching and learning during pandemics. Therefore, this study explores how Foundation Phase lecturers and student teachers coped with teaching, learning, and assessing Mathematics during and after COVID-19.

## Teaching and learning of mathematics during the COVID-19 pandemic

Since Mathematics has its unique language and the process of creating mathematical knowledge has its specifics, for example, according to Vukovic and Lesaux (2013), self-study of Mathematics (from a textbook or other materials) is very difficult for most students. According to Schleppegrell (2007), Mathematics, more than any other scientific discipline, depends on the emphasis on spoken language, the teacher's explanation, and social interactions with other learners. In the exposure stage, this interaction between the lecturer and the students and the students themselves seems irreplaceable. Much mathematical knowledge arises from mutual discussion in pairs or groups. The lecturer provides the students with mathematical content and a formal context of the procedures used, such as the notation in the division process by two and more digit numbers or the geometrical construction. Many of these specifics were technically unattainable despite teachers' efforts to improve information technology skills.

## Challenges faced in teaching mathematics during the COVID-19 crisis in higher education

The temporary transition from their regular teaching delivery methods compelled them to swiftly transition to what Hodges et al. (2020) characterized as emergency remote teaching. There are many challenges that lecturers and students faced during the crisis in higher education. In this chapter, the following challenges were outlined. The field of Mathematics frequently calls for the utilization of manipulatives, textbooks, and various other physical resources to facilitate learning. When educating students online, providing them with the necessary materials is challenging. Lecturers must be flexible and creative to show new ideas through alternate methods or provide students with digital tools that can stand in for traditional hands-on materials.

Barriers posed by technology during teaching and learning include inadequate training, insufficient technical help, lack of peer support, slow internet connectivity, frequent power failures, inadequate software quality, substandard hardware quality, unavailability of software, insufficient knowledge, and a lack of confidence were arrayed as problems in teaching Mathematics. Transitioning to online platforms and digital technologies can be difficult for students and lecturers. During the crisis, not every student had the same level of access to dependable internet connections, devices, or software that was important for their education. Problems with technology, such as interruptions in the audio or video signal, can make the process of teaching and learning much more difficult. In addition, not all mathematical ideas can be effectively communicated through virtual means, particularly those dealing with manipulating physical objects or activities that need direct participation from the students. Online teaching must be intelligently combined with face-to-face teaching, and distance education needs to be skillfully blended to get the best possible educational outcomes (El Firdoussi et al., 2020).

Student motivation and discipline may suffer due to the change to online learning since the student's home environment may provide a different structure and accountability than the traditional classroom setting. During virtual lessons, students may have trouble managing their time effectively, become easily distracted, or struggle to maintain attention to the presented material. It becomes a huge problem to keep students interested and involved in the mathematical teachings they are receiving. Interaction and engagement are limited in remote learning, which frequently does not provide the face-to-face interaction necessary for teaching Mathematics. When a teacher is not physically present in the classroom, it can be difficult to maintain students' attention and engage them in learning activities. It may be difficult to assess the level of students' comprehension and adequately respond to the questions they pose when there is no opportunity for instant feedback and no human encounters. Hofer et al. (2021) aver that lecturers and student teachers were unprepared for the shift from traditional classroom settings to online learning, which occurred seemingly overnight.

Evaluating the development of students and giving them prompt feedback can be more difficult in remote settings. It is possible that more conventional forms of testing, including in-person quizzes or examinations, will need to be updated or even phased out entirely in favor of alternate evaluation strategies. Individualized support and varying rates of educational progression are two aspects that can make it difficult to meet the varied educational requirements and tempos of students in an online environment. Some students may need additional support or accommodations, which can be supplied more readily in person. During virtual sessions, responding to individual students' specific enquiries or confusions can be time-consuming and inefficient, making individualized education difficult to achieve. The other challenges that mathematics lecturers and student teachers encounter while teaching mathematics on campus are inadequate resources, commitment, internet access, complementary approaches to face-to-face, online pedagogies, and technical skills. Therefore, existing online Mathematics education guidelines do not address Mathematics self-concept (Lee and Kung, 2018).

The socio-economic environment in Africa has a significant impact on digital learning, particularly in the field of mathematics education, both during and after times of crisis. It is essential to have a solid understanding of these factors to establish effective educational practices that address inequities and harness opportunities to improve learning outcomes. During times of crisis during the COVID-19 pandemic, socio-economic elements impacted digital learning in mathematics education within the framework of higher education. The socio-economic aspects influence digital learning and access to technology. Students from lower-income families frequently do not have access to critical digital tools like laptops, tablets, and internet connections. According to Afolabi (2021), the digital divide makes it substantially more difficult for them to participate in online learning communities. Adedoyin and Soykan (2020) concur that rural areas often have a lower internet access infrastructure than metropolitan centers, contributing to the worsening educational disparities.

## Designing the mathematics class activities using blended pedagogies for the foundation phase level

Designing Mathematics class activities using blended pedagogies at the Foundation Phase level can significantly enhance teaching, learning, and assessment outcomes. Blended pedagogies combine traditional face-to-face teaching with digital technology and online resources to create a more interactive and engaging learning experience. First, it is crucial to balance face-to-face and online interactions, and lecturers should balance face-to-face interactions and online activities. While online resources can enhance learning, face-to-face interactions are essential for building relationships and addressing individual student needs. Second, there is a need for continuous improvement, regular evaluation of the effectiveness of the blended learning approach, and openness to feedback from both students and lecturers. Third, using visuals and infographics to incorporate visuals, infographics, and data representations enhances students' understanding of mathematical concepts. Visual aids are powerful tools for visual learners and can make complex information accessible. Lastly, lecturers should encourage collaborative learning through group activities, discussions, and peer teaching. Blended learning environments provide opportunities for students to work together both in the physical classroom and through online platforms. One strategy is using gamification elements like quizzes, puzzles, and interactive games to make learning Mathematics more enjoyable and engaging. Gamified activities can motivate students to participate actively in the learning process.

By making use of interactive web resources, the lecturers can develop learning experiences that are both interesting and engaging. Teachers can use virtual manipulatives, including virtual base-ten blocks, mathematical objects, and forms, to assist pupils in comprehending abstract mathematical ideas. The educational games can incorporate mathematical games that turn learning into an enjoyable and engaging experience while offering practice in essential abilities. Tools such as videos and tutorials can provide access to instructional videos that teach modules in a way that is interesting and easy to understand. In addition, lecturers can build immersive and hands-on learning through physical manipulatives. These manipulatives include tangible objects such as counting beads, shape sorters, and measuring tools used in face-to-face sessions. Developing activities that help students see the relevance and application of their learning has become increasingly crucial. These activities should be designed to connect mathematical concepts to real-world problems.

The above are the strategies for mathematics class activities, and the lecturer training program must be flexible. Some strategies include offering additional support to struggling students, providing technology resources, re-evaluating assessment criteria, and incorporating a mix of assessment methods to comprehensively assess student teachers' capabilities. As we move beyond the pandemic, we must continue reflecting on the lessons learnt and refining assessment approaches for student teachers in a rapidly changing educational landscape. Therefore, blended pedagogies integrate traditional face-to-face teaching methods with online digital technologies to create a dynamic and interactive learning environment. Blended pedagogies are often used during the Foundation Phase of education, which typically encompasses early infancy to early primary education.

## Improvement of teaching mathematics after the COVID-crisis in higher education

The COVID-19 pandemic has significantly impacted education systems worldwide, including the teaching and learning of Mathematics. However, it has also presented opportunities for improvement and innovation in teaching Mathematics.

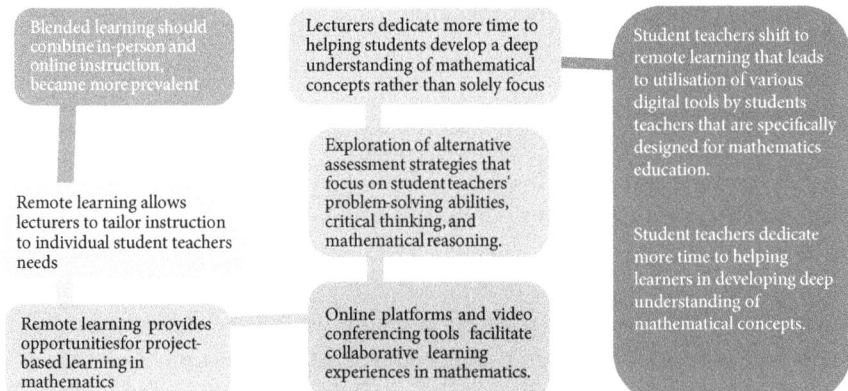

**Figure 11.1:** The improvement of teaching mathematics after a pandemic crisis in higher education.

Figure 11.1 illustrates some potential improvements in teaching Mathematics after the COVID-19 crisis. The COVID-19 pandemic has caused disruptions in traditional educational methods and accelerated notable enhancements in mathematics teaching at the university level. Amidst the epidemic, institutions and lecturers have made adjustments to overcome the difficulties, resulting in several improvements and ideas that could significantly reshape mathematics education in the aftermath of the crisis. The significant advancements in mathematics instruction in higher education after the COVID-19 pandemic concentrate on integrating technology, innovative teaching methods, enhancing skills and knowledge, and formulating policies.

## Teaching method used in teaching mathematics during and after the crisis

Teaching Mathematics during and after the COVID-19 pandemic requires adaptation to accommodate remote learning and address the unique challenges students face. The following are the teaching methods the lecturers use during teaching and learning.

The teaching methods used during a crisis:
· Online teaching method through Teams;
· Recorded lessons and sent through WhatsApp;
· Moodle platform and email;
· Online quizzes.

The teaching method used after the crisis:
- Blended learning;
- Face to Face;
- Moodle platform and email;
- Synchronous and Asynchronous learning.

Therefore, the post-COVID-19 era necessitates the implementation of groundbreaking modifications in mathematics education to guarantee the subject's continued viability and relevance in a constantly evolving world. With the implementation of this powerful teaching method, we will be able to establish a mathematics education system that is both robust and dynamic. Not only can these teaching methods improve students' mathematical abilities and conceptual comprehension, but they also provide them with the critical skills necessary to confront difficulties they will face in the real world and become active contributors to the sustainable development of society.

## Theoretical Framework

This chapter is underpinned by Venkatesh, Thong, and Xu's 8-point 2003 model on the Extended Unified Theory of Acceptance and Use of Technology (UTAUT), which predicts behavioral acceptance of technological innovations (Venkatesh et al., 2003). The UTAUT is a well-known model that explains the factors influencing the acceptance and use of technology. When teaching Mathematics through online platforms in higher education, an extended version of the UTAUT can be developed to address the specific factors and challenges associated with this domain. The theoretical model of UTAUT suggests that the actual use of technology is determined by behavioral intention. The perceived likelihood of adopting the technology depends on the direct effect of four key constructs: performance expectancy, effort expectancy, social influence, and facilitating conditions. The effect of predictors is moderated by age, gender, experience, and voluntariness of use (Venkatesh et al., 2003).

UTAUT has made several important contributions to the body of knowledge in education. The model compares popular technology acceptance ideas, which frequently offer contradictory or partial viewpoints, to provide empirical insight into the technology acceptance process. UTAUT reveals that proposed components account for seventy per cent of the variance in use intention (Venkatesh et al., 2003). This model has more substantial predictive power than the other models investigating technology adoption, such as Davis (1993) and Sheppard et al. (1988). The fact that certain constructions have an interactive influence on personal and demographic aspects indicates the complexity of the process of technology acceptance, which is reliant on the ages, genders, and levels of experience of individual users (Venkatesh et al., 2003; Marikyan and Papagiannidis, 2023). Thus, this theory enhances the understanding of how the Foundation Phase student teachers and their lecturers coped with teaching, learning, and assessing Mathematics during and after the COVID-19 pandemic, which forced the lockdown of all socio-economic activities, including education. By considering these extended factors within the UTAUT framework in this chapter, researchers and lecturers can gain insights into the acceptance and use of technology for teaching Mathematics online in higher education (Venkatesh and Davis, 2000). This can help guide the development of strategies and interventions to promote adequate integration and adoption of online platforms in mathematics education.

## Research Design and Methods

An explorative research design was used to gain a better knowledge of the phenomenon that was the focus of the investigation, and a qualitative study was conducted (Lewis and Thornhill, 2012). The exploratory design is used because it is the success of interpretivism and depends on one's ability to understand subjective meaning. It states that meaningful reality is produced via interacting with persons and their surroundings. Three steps include identifying the problem, conceptualizing data collection, and data analysis methods, and collecting and analyzing data to determine the preliminary results (George, 2023). As a result, this qualitative study aimed to explore how Foundation Phase lecturers and student teachers coped with the teaching, learning, and assessment of Mathematics during and after COVID-19, and the viewpoints that the participants themselves ascribed to the social issue being studied.

An interpretive paradigm and a qualitative technique to explore teaching, learning, and mathematics assessment during and after the crisis in higher education were used to gather the data. The interpretive paradigm was used to get reliable information from the participants. The purposeful sampling was used to select ten student teachers who registered for the Mathematics module in the Bachelor of Education in Foundation Phase Teaching (BEd FP Teaching) program. Three lecturers who taught Mathematics modules during the COVID-19 pandemic in the BEd FP Teaching were also selected. Etikan et al. (2016) define the purposive sampling technique as the deliberate participant selection based on the participant's attributes. They would reflect and share their lived practical experiences. Focus group interviews were conducted to get reliable information from the student teachers. The three lecturers were individually interviewed and observed while teaching Mathematics physically after a crisis. The sample size, therefore, consisted of thirteen participants, three lecturers, and 10 student teachers. Participants operated and were seen from their natural surroundings because the interpretative study aimed to understand them and their engagement with the world to make sense of their lived world experiences (Gravetter and Forzano, 2016).

Inductive thematic analysis was conducted on the gathered data, which was primarily narrative. Inductive thematic analysis was used because themes, codes, and categories emerged from the data. It was decided to use Creswell's (2014) suggested thematic analysis method. The Creswell thematic analysis is used because it is a qualitative research method that involves identifying, analyzing, and interpreting patterns and themes in data. This flexible approach can be applied to various data types, including interviews, focus groups, and written documents. The content of the discussion was coded to discover recurring or underlying themes within the solicited data and throughout the debate. Chowdhury (2014) sees the relevance of interpretivism because it enables the researcher to see the world through the eyes of the people being studied and allows them numerous perspectives of reality.

## Ethical Considerations

Ethical clearance was obtained from the university's higher degree committee, where the study was conducted. As required in research with humans and animals, the ethical clearance number is FHSSE/23/ECE/01/0601. The rights of the participants, such as the right to secrecy, the right to privacy and dignity, the right to voluntary participation, and the right to withdraw from the process, were upheld throughout the process (Holland, 2019).

## Research Findings

Four themes emerged from the data analysis, and the findings are discussed below. The participants were coded as follows: LA to LC, which stands for lecturers A to C. ST 110, which stands for student teachers 1–10.

### Theme 1: Challenges of teaching mathematics using blended learning and hybrid models

Lecturers reported that teaching Mathematics during crises was difficult for them as most of the student teachers are from disadvantaged areas, and the network was challenging. Teaching resources were a big problem. Hybrid learning's online components are intended to replace in-person instruction without sacrificing its allure and potency. The synchronous nature of hybrid learning allows remote students to communicate with the instructor and the other students in the class while still having the option to do so. This means that in higher education institutions where the virus hit hardest, lecturers taught student teachers online full-time as soon as possible.

Participant LB said:

*But once we knew more about how COVID worked and spread, the university switched to a "hybrid" approach, where some student teachers came to class in person and others called in using Microsoft Teams.*

Student teachers (ST) reported that learning mathematics during crises was challenging as the module needed demonstration and different manipulatives. It was pointed out that for the lecturers to demonstrate a Mathematics lesson online, sharing on WhatsApp made it difficult for the students to understand the instructions.

Participant ST4 shared that:

*It was not simple for the lecturer to demonstrate how I should teach patterns in the Foundation Phase classroom through WhatsApp because it was not easy to share the various patterns.*

Similarly, ST5 echoed the following:

*Some of us did not receive any communication from lecturers as we do not have smartphones and WIFI to receive the messages.*

## Theme 2: Assessing student teachers' learning during and after the pandemic posed unique challenges

Some student teachers revealed that assessment during crises was good because some lecturers gave them take-home tests and examinations. Other student teachers reported getting bored with being assessed by taking home tests and examinations because they did not gain knowledge and could not master the mathematics content. They were coping with the answers from the books. According to Rich (2011), students examine their notes and textbooks more frequently while preparing for a take-home test than when preparing for an in-class test. Students are also more inclined to raise questions in class and paraphrase the lesson on their own terms.

In our study, all the lecturers reported that the assessment was very confusing and frustrating because they were instructed to send the tests and examinations to the student teachers so that they could write and send the scanned answer book.

They even indicated that some student teachers sent unclear scripts as they were using their phones to scan the scripts. According to Haynie (2003), students only focus on the knowledge necessary to finish the take-home examination. They typically do not acquire additional understanding from other learning materials. Students will be better prepared for this kind of test if the questions and themes on the examination are more complicated, and the questions and topics on the examination should also be more difficult.

Participant L2 also added that:

*Take-home tests and exams influence students to cheat, and correcting the scripts is difficult because they copy and paste information as it is.*

## Theme 3: Inadequate teaching resources in mathematics during and after the COVID-19 crisis

Student teachers reported that Mathematics lectures were poorly resourced. The COVID-19 crisis from 2020 to the beginning of 2022 affected their learning as they were not provided with enough data that covered all registered modules by Foundation Phase student teachers. Some students concurred that they had no laptops or smartphones for use in attending online classes. During focus group interviews, [ST8] also reported that:

*It was not easy to understand the content when lecturers taught Mathematics without demonstrating and explaining Mathematics when there was no face-to-face interaction.*

[ST4] added that:

*It is not simple to keep a record of student homework and assignments in real-time, and how to evaluate their mathematical understanding at the end of the lesson.*

Brodie et al. (2021) supported that using WhatsApp to teach mathematics is incredibly challenging. This left us feeling depressed about the amount of class time our students' teachers were missing and the potential for previously existing disparities to become even more pronounced amongst educational institutions.

[LA] asserted that:

*Crisis prolonged school irregular attendance among student teachers, leading to learning gaps and knowledge loss among students.*

[LC] supported that

*Inadequate teaching resources during this period exacerbate the problem and make it challenging for students to catch up on missed material when the university reopens.*

Student teachers interacted with their lecturers during observation when lecturers A, B, and 3C were busy teaching face-to-face after the crisis. Demonstrations were made practically as the student teachers learned how to use and solve problems and build a sense of measurement that increases their spatial awareness when teaching mathematics. All these experiences were lacking in schools during COVID-19. According to Glass and Sue (2008), a demonstration of assisting student teachers in developing Mathematics teaching materials, such as number charts and shapes, was given as a project. It was the most desired learning object that had the most significant impact on the face-to-face learning process.

This was also reported by ST5 as follows:

*During the crisis, they did not get an opportunity to develop their Mathematics teaching material that they could use during teaching practice.*

During the crisis, lecturers reported that they were inflexible and that the curriculum was only delivered online. However, after the crisis, they could choose either the curriculum's online or contact delivery mode.

LB concurred that:

*After the crisis, they are flexible and adapted various teaching methods to enhance understanding of mathematics teaching concepts.*

ST3 also reported that:

*After the COVID-19 pandemic, the demonstration and practical lesson on how to instil Mathematics skills in learners in schools would improve for me.*

## Discussion

In this chapter, the findings showed that blended learning and hybrid models during the crisis did not enhance student engagement because the delivery mode of the curriculum was online only, as the country was in total lockdown (Singh et al., 2021). Lecturers gave student teachers assignments through WhatsApp. The student teachers were unmotivated and lost interest in the mathematics modules. The deeper discussions addressed specific challenges and fostered a stronger teacher-student relationship, but were minimal. The findings indicated that the crisis often disrupted access to technology, such as computers, tablets, and the internet, on the campus. Without these tools, student teachers struggled to provide digital resources, attend online lessons, or interact with the e-learning platforms that could enhance students' understanding of mathematical concepts. Not all student teachers had equal access to technology or reliable internet connections during the pandemic. This digital divide hindered their participation in online learning and assessments, potentially creating disparities in the evaluation process. Sumardi and Nugrahani (2020) supported that due to the lack of knowledge about online teaching and the sudden shift from traditional on-campus teaching to emergency remote teaching, lecturers implemented teaching strategies designed initially for face-to-face instruction, which presented new challenges for the student teachers.

This lack of tangible materials can hinder students' ability to practice and apply mathematical principles. The findings also showed that lecturers face increased workloads and stress, making it difficult to focus and deliver effective Mathematics lessons. Besides, the distribution of textbooks and other physical teaching materials was disrupted, leaving students without essential resources to study Mathematics.

Student teachers typically benefit from in-person teaching practice (Blane, 2022), which was severely impacted during the pandemic due to school closures and restrictions. The lack of sufficient practical experience could affect their development as lecturers and make it challenging to assess their teaching skills accurately. Conducting assessments in virtual settings might not always mirror real-life teaching scenarios. It can be difficult to gauge how well student teachers would perform in traditional classrooms based solely on remote observations or virtual lessons. Ensuring the authenticity of student teachers' work during remote assessments posed a challenge. Also, without direct supervision, there was a higher risk of academic dishonesty or using unauthorized resources.

After the crisis in higher education, teaching Mathematics using blended learning and hybrid models has the potential to revolutionize traditional classroom dynamics, providing students with more personalized learning experiences. Holland (2022) concurs that blended learning can enhance academic achievements in mathematics among higher education students. The primary focal point of this approach lies in considering factors that effectively engage, assist, and foster the development of students' comprehension of mathematical concepts. Therefore, engaging and equipping lecturers' learning experiences while arming lecturers with valuable data-driven information enhances their effective teaching. However, successful implementation requires careful planning, ongoing support, and a commitment to addressing the associated challenges.

## Conclusion

From the discussion, the chapter concludes that the three themes reflect the transformative impact of the COVID-19 pandemic on mathematics education, as the period required diverse teaching, learning, and assessment approaches that departed from the traditional ones. While the challenges were significant, they also presented opportunities for innovation and growth in teaching practices, fostering resilience and adaptability among lecturers. On the contrary, the COVID-19 pandemic posed unique and unprecedented challenges to student teachers' learning experiences. The sudden shift to remote or hybrid learning disrupted traditional teaching practices and required significant adjustments from lecturers and students. Not all students had equal access to technology or stable internet connections, which hindered their ability to engage in online learning activities fully. Many student teachers demonstrated remarkable adaptability and resilience in navigating the changes brought about by the pandemic.

The assessment of student teachers' learning during and after the pandemic highlighted the need for flexibility, adaptability, and supportive measures to ensure effective teacher education. The pandemic emphasized the importance of technology integration, personalized support, and mental health considerations in preparing future teachers for the changing education landscape. Future research possibilities emphasize that lecturers should continue to use blended learning to enhance academic achievements in higher education by directing attention to these potential areas of future research. The stakeholders in the field of Mathematics Education, including lecturers, policymakers, and academics, might engage in cooperative efforts to construct a more resilient and flexible framework for Mathematics education. This framework would be capable of effectively addressing and adapting to crises within higher education.

## Acknowledgments

The researchers acknowledge the University of Venda Higher Degree Ethical Committee for reviewing research projects and granting the Ethical clearance certificate.

## References

Adedoyin, O. B., & Soykan, E. (2020). COVID-19 pandemic and online learning: The challenges and opportunities. *Interactive Learning Environments. Advance Online Publication.* <https://doi.org/10.1080/10494820.2020.1813180>.

Afolabi, F. (2021). Mobile learning as a catalyst for inclusive education in Africa. *International Journal of Educational Development in Africa* 8(1), 45–59. <https://doi.org/10.1080/17533171.2021.1874972>.

Aldon, G., Cusi, A., Schacht, F. & Swidan, O. (2021). Mathematics in the context of lockdown: A study focused on teachers' Praxeologies. *Educational Science 11*(2), 38. <https://doi.org/10.3390/educsci11020038>.

Blane, P. (2022). *The Benefits of In-person Learning.* Retrieved from <https://cbassociatetraining.co.uk/the-benefits-of-in-person-learning/>.

Brodie, K., Gopal, D., Moodliar, J., & Siala, T. (2021). Bridging powerful knowledge and lived experience: Challenges in teaching mathematics through COVID-19. *Pythagoras 42*(1), a593. <https://doi.org/10.4102/pythagoras. v42i1.593>.

Chauhan, S. (2017). A meta-analysis of the impact of technology on learning effectiveness of elementary students. *Computers & Education 105*, 14–30. <https://doi.org/10.1016/j.compedu.2016.11.005>.

Chowdhury, M. F. (2014). Interpretivism in aiding our understanding of the contemporary social world. *Open Journal of Philosophy 4*(03), 432–438. <https://doi.org/10.4236/ojpp.2014.43047>.

Creswell, J. W. (2014). *A Concise Introduction to Mixed Methods Research.* SAGE publications.

Dhawan, S. (2020). Online learning: A panacea in the time of COVID-19 crisis. *Journal of Educational Technology Systems 49*(1), 5–22. <https://doi.org/10.1177/0047239520934018>.

Drijvers, P. (2020). *Moving Forward in the Midst of a Pandemic: International Lessons for Math Teachers.* Presentation at the National Academies of Sciences, Engineering, and Medicine. Available online: <https://www.nationalacademies.org/event/07-09-2020/math-distance-distance-mathematics-teaching-during-covid-19-lockdown>.

El Firdoussi, S., Lachgar, M., Kabaili, H., Rochdi, A., Goujdami, D., & El Firdoussi, L. (2020). Assessing distance learning in higher education during the COVID-19 pandemic. *Education Research International*, 1–13. <https://doi.org/10.1155/2020/8890633>.

Etikan, I., Musa, S. A., & Alkassim, R. S. (2016). Comparison of convenience sampling and purposive sampling. *American Journal of Theoretical and Applied Statistics 5*(1), 1–4. <https://doi: 10.11648/j.ajtas.20160501.11>.

George, T., (2023). Exploratory research | definition, guide, & examples. *Scribbr.* Retrieved from <https://www.scribbr.com/methodology/exploratory-research/>.

Glass, J., & Sue, V. (2008). Student preferences, satisfaction, and perceived learning in an online mathematics class. *MERLOT Journal of Online Learning and Teaching 4*(3), 325–338.

Godber, K. A., & Atkins, D. R. (2021). COVID-19 impacts on teaching and learning: A collaborative autoethnography by two higher education lecturers. In *Frontiers in Education*, (Vol. 6, p. 647524). Frontiers Media SA. <https://doi.org/10.3389/feduc.2021.647524>.

Govindarajan, V., & Srivastava, A. (2020). What the shift to virtual learning could mean for the future of higher ed. *Harvard Business Review 31*(1), 3–8. <https://www.accs.edu/wp-content/uploads/2020/06/What-the-Shift-to-Virtual-Learning-Could-Mean-for-the-Future-of-Higher-Ed.pdf>.

Gravetter, F. J., & Forzano, L.-A. B. (2016). *Research Methods for the Behavioral Sciences*. 5th edn., Cengage: Stamford.

Haynie III, W. J. (2003). Effects of take-home tests and study questions on retention learning in technology education. *Journal of Technology Education 14*(2), 6–18. <https://er.educause.edu/articles/2020/3/the-difference-between-emergency-remote-teaching-and-online-learning>.

Hofer, S.I., Nistor, N., & Scheibenzuber, C. (2021). Online teaching and learning in higher education: Lessons learned in crisis situation situations. *Computers in Human Behavior 121*, 106789; *Journal of Distance Education 22*(2), 81–93. <https://doi.org/10.1016/j.chb.2021.106789>.

Holland, N. (2022). *Using Blended Learning to Strengthen Math Skills*. The United States of America: Lucas Educational Foundation.

Huber, S. G., Helm, C. (2020). COVID-19 and schooling: evaluation, assessment and accountability in times of crisis situation—reacting quickly to explore key issues for policy, practice and research with the school barometer. *Educational Assessment, Evaluation and Accountability 32*, 237–270. <https://doi.org/10.1007/s11092-020-09322-y>.

Krishnamurthy, S. (2020). The future of business education: A commentary in the shadow of the Covid-19 pandemic. *Journal of Business Research 117*, 1–5. <https://doi.org/10.1016/j.jbusres.2020.05.034>.

Lee, C.-Y., & Kung, H.-Y. (2018). Math self-concept and mathematics achievement: Examining gender variation and reciprocal relations among junior high school students in Taiwan. *EURASIA Journal of Mathematics, Science and Technology Education 14*(4), 1239–1252. <https://doi.org/10.29333/ejmste/82535>.

Lewis, P., & Thornhill, A. (2012). *Research Methods for Business Students*, 6th edn, Pearson Education Limited.

Li, Q., & Ma, X. (2010). A meta-analysis of the effects of computer technology on school students' mathematics learning. *Educational Psychology Review 22*, 215–243. <https://doi.org/10.1007/s10648-010-9125-8>.

Marikyan, D., & Papagiannidis, S. (2023). Unified theory of acceptance and use of technology: A review. In S. Papagiannidis (Ed.), *Theory Hub Book*. Available at <https://open.ncl.ac.uk>/ISBN: <9781739604400>.

Naidoo, J. (2020). Postgraduate mathematics education students' experiences of using digital platforms for learning within the COVID-19 pandemic era. *Pythagoras 41*, a568. <https://doi.org/10.4102/pythagoras.v41i1.568>.

Olivier, W. (2020). Education post-COVID-19: Customised blended learning is urgently needed. Available online: rates between take-home and in-class examinations. *International Journal of University Teaching and Faculty Development 2*, 1–10.

Pokhrel, S., & Chhetri, R. (2021). A literature review on impact of COVID-19 pandemic on teaching and learning. *Higher Education for the Future 8*(1), 133–141. <https://doi.org/10.1177/2347631120983481>.

Rich, J. (2011). *An Experimental Study of Differences in Study Habits and Long-Term Retention Strategy for Pre-service Language Teachers Amid COVID-19 Pandemic*. Turkish Online.

Schleppegrell, M. J. (2007). The linguistic challenges of mathematics teaching and learning: A research review. *Reading & Writing Quarterly 23*(2), 139–159. <https://doi.org/10.1080/10573560601158461>.

Singh, J., Steele, K., & Singh, L. (2021). Combining the best of online and face-to-face learning: Hybrid and blended learning approach for COVID-19, post vaccine, & post-pandemic world. *Journal of Educational Technology Systems 50*(2), 140–171. <https://doi.org/10.1177/00472395211047865>.

Sumardi, S., & Nugrahani, D. (2021). Adaptation to emergency remote teaching: Pedagogical technology education. *Journal of Technology Education 14*(2), 6–18. <https://doi.org/10.17718/tojde.906553>.

Venkatesh, V., & Davis, F. D. (2000). A theoretical extension of the technology acceptance model: Four longitudinal field studies. *Management Science 46*(2), 186–204. <https://doi.org/10.1287/mnsc.46.2.186.11926>.

Venkatesh, V., Morris, M. G., Davis, G. B., 7 Davis, F. D. (2003). User acceptance of information technology: Toward a unified view. *MIS Quarterly*, 425–478.

Vukovic, R. K., & Lesaux, N. K. (2013). The language of mathematics: Investigating the ways language counts for children's mathematical development. *Journal of Experimental Child Psychology 115*(2), 227–244. <https://doi.org/10.1016/j.jecp.2013.02.002>.

CHAPTER 12

# Summing Up: Digitalization in STEM Education in Post-Pandemic Africa

*Brantina Chirinda[1] and Jayaluxmi Naidoo[2]*
[1]University of Johannesburg
[2]University of KwaZulu-Natal

## Introduction

The COVID-19 pandemic has impacted science, technology, engineering, and mathematics (STEM) education and hastened digital transformation across Africa. As we navigated our transformed education environments, several key learnings emerged that will advance our future efforts to scaffold STEM pedagogy. Hence, we embarked on the journey of conceptualizing this edited book. The purpose of this chapter is to conclude this edited book. We accomplish this by revisiting the purposes and scope for starting our conceptualization of this edited book. We then highlight and summarize key points showcased in this edited book. After that, we examine the themes that have emerged from the different chapters and research pieces. Additionally, we provide our final thoughts, reflections, implications, and future directions. Subsequently, we conclude the chapter.

## The Purpose and Scope of the Book

The digital transformation in STEM education in Africa is an important and timely topic. This edited book, *STEM Education in the Post-pandemic Learning Space: Digitilization in Africa*, underscores the impact of the COVID-19 pandemic on STEM education in Africa and the potential of digitalization to enhance the quality and accessibility of STEM learning post-pandemic. In the wake of the pandemic, digitalization has become an urgent necessity for STEM education in Africa (Mhlanga et al., 2022). As African nations adapt to the post-pandemic learning landscape, there is an undeniable emphasis on harnessing digital technologies to elevate STEM education. The shift to digital learning is essential for equipping students with the skills required for the future workforce, particularly in STEM. By leveraging digital transformation, African countries can bridge the educational gap and foster innovation and growth in the field of STEM (Maarman, 2023). This transformation is crucial for overcoming the challenges of accessing quality education and resources that have long persisted across the continent.

## Summary of Key Points

This book delves into the challenges confronted by STEM teachers and teacher educators during the abrupt transition to online teaching, prompting inquiries into the insights gained and strategies to incorporate the advantages of online instruction into physical classrooms. This book emphasizes the exacerbation of digital inequalities in STEM education caused by the pandemic. The digitalization of STEM education in Africa presents unprecedented opportunities and formidable challenges. While it holds the potential to reach a broader and more diverse student demographic, including those in remote or marginalized areas, the digital divide remains a significant obstacle. Bridging this gap requires comprehensive infrastructure development, substantial investment in technology, and policy reforms in contexts of disadvantage to ensure equitable access to digital learning resources. This entails initiatives to amplify access to online resources, develop robust digital learning platforms, and empower educators with the necessary skills for effective digital teaching methodologies. Moreover, successfully integrating digital technologies in STEM education demands rigorous teacher training and ongoing professional development. Educators must be adept at leveraging digital tools, adapting to online teaching environments, and facilitating engaging and collaborative learning experiences for their students.

This book foregrounds that the digitalization of STEM education in the post-pandemic era in Africa promises to revolutionize educational opportunities and prepare the next generation for an increasingly technology-driven world. However, it is imperative to confront the existing challenges and inequities to guarantee that all students have equal access to high-quality STEM education in the digital age.

Chapter 2 discusses the importance of a relevant STEM curriculum in Africa to address societal challenges and support sustainable development. It emphasizes the need for curriculum reorientation and the incorporation of transformative pedagogies. The African Union's Agenda 2063 prioritizes science, technology, and innovation to support sustainable development, making STEM education crucial for Africa's development agenda. Overall, it calls for a harmonized STEM curriculum to develop the human capital needed to drive Africa's sustainable development.

In Chapter 3, the impact of the COVID-19 pandemic on STEM education in Sub-Saharan Africa is discussed, highlighting the challenges teachers face, the significance of digital tools, and the inequalities in technology access. The chapter also presents original research findings from Namibia, South Africa, Zambia, and Zimbabwe and explores factors influencing teachers' acceptance of digital transformation in STEM education using the UTAUT2 model. The need to rethink and re-imagine STEM education in Africa, despite the digital divide, is emphasized.

Chapter 4 delves into the impact of virtual learning spaces on postgraduate mathematics education in Egypt, focusing on female students. It challenges the belief that in-person instruction is always superior and emphasizes the inclusive nature of online learning, particularly for marginalized students. Based on interviews with eight postgraduate students facing mobility challenges, the research highlights the potential of virtual learning for promoting inclusivity. The study recommends modifying the current classroom format to be more inclusive and advocates for greater inclusion of postgraduate female students in Africa. The study aims to advocate for postgraduate female learners in Egypt and reflects on how the shift to virtual learning during the COVID era has created opportunities for inclusion. It also delves into the Egyptian context of Higher Education, presenting statistics related to the selected participant sample. Situated within an inclusive pedagogy framework, the study outlines the research question and findings, culminating in a discussion of the implications of the study.

Chapter 5 discusses postgraduate students' experiences with digital pedagogy for mathematics education in South Africa and highlights the use of digital tools and platforms during the COVID-19 pandemic. It emphasizes the strengths, challenges, and implications of digital pedagogy, stressing the importance of preparation workshops, access to digital resources, and active engagement for effective teaching and learning. Additionally, it touches upon the role of digital pedagogy in the Fourth Industrial Revolution and its impact on mathematics education, emphasizing the need for critical thinking, collaboration, and real-world relevance in teaching mathematics.

Chapter 6 examines the experiences of teachers in STEM academic centers in Mpumalanga, South Africa, before, during, and after the COVID-19 pandemic. It underscores the pandemic's impact on education, the transition to remote teaching and learning, and the challenges and opportunities of integrating e-learning in mathematics education, specifically focusing on teachers in rural areas.

Chapter 7 examines the experiences of postgraduate STEM students with online teaching and learning during the COVID-19 pandemic, particularly in South Africa. It explores the challenges STEM educators face in online education and the potential benefits of using ICT to help students develop a deeper understanding of scientific concepts across disciplines. The chapter stresses the importance of incorporating technology in STEM education and discusses how technology can help overcome language barriers in STEM teaching. Furthermore, the chapter underscores the need for STEM educators to have strong content knowledge and online teaching skills to foster key competencies in students and advocates for an integrated approach to teaching the four STEM disciplines.

Chapter 8 examines technology integration into mathematics education at a South African university. It emphasizes preparing students for the digital workplace and discusses the effect of the COVID-19 pandemic on the shift to online teaching. The study recommends technology training for mathematics educators in rural universities and underscores the importance of user-friendly technology in mathematics education programs. Additionally, the literature review covers the effects of the COVID-19 pandemic on traditional classroom learning and the need for alternative strategies.

Chapter 9 examines STEM teaching and learning in early childhood classrooms in Zimbabwe during and after the COVID-19 pandemic. It discusses the challenges teachers and students face, recommends implementing long-term educational policies, and suggests training for early childhood educators to improve their STEM expertise. Additionally, it explores the shift to online education and the use of digital tools and platforms by ECD teachers during and after the COVID-19 lockdowns.

Chapter 10 examines the impact of the COVID-19 pandemic on STEM education and the challenges teacher educators face in adapting to the new normal. It delves into the challenges educators encounter when transitioning to online platforms and offers strategies for integrating digital and face-to-face teaching. The study recommends training programs for educators to enhance their skills and competencies in response to potential educational system changes. Additionally, it highlights the digital divide between rural and urban areas, as well as between private and public institutions, particularly in developing countries like Tanzania.

Chapter 11 discusses the challenges higher education institutions face in Africa, particularly in mathematics teaching, during and after the COVID-19 pandemic. It underscores the impact of the pandemic on teaching, learning, and assessment of mathematics, as well as the shift to online education. The chapter emphasizes the need for blended pedagogies and the utilization of technology in mathematics education, more specifically, in the context of the fourth industrial revolution. Additionally, it provides insights into the experiences and challenges encountered by lecturers and student teachers in Africa, such as inadequate resources and unreliable internet service.

## Final Thoughts and Reflections

The COVID-19 pandemic has been a transformative period for education worldwide (Bozkurt et al., 2020; Sedaghatjou et al., 2023), and Africa is no exception (Badat, 2020; Kolog et al., 2022; Mhlanga et al., 2022; Naidoo, 2020). As STEM lecturers in Africa, we have personally experienced the challenges and opportunities that digitalization brings to educational environments. For example, in the early months of the pandemic, the universities where we lecture switched to online learning swiftly. This transition was both exciting and intimidating. We remember our first Zoom lectures; although it was an introductory lecture and the students were trying their best to engage with the content being discussed, we were often interrupted due to poor internet connectivity or family members of students wanting to participate in the lecture. We traversed these uncertain times with a mixture of hope and despair. Other research also supports these sentiments (Caldararo, 2020; Graham et al., 2022; Knowles et al., 2022). Now, post-COVID-19, we are accustomed to successfully navigating our digitalized educational environments. We have gained much experience and learned valuable lessons during the pandemic to support us as we embrace digitalization in Africa post-COVID-19.

Our experiences of navigating digitalized STEM education in Africa were not always positive due to the lack of infrastructure, limited internet connectivity, inadequate access to the necessary devices and tools, and socio-economic disparities that continue in our communities. Our colleagues have tried their best to overcome these challenges. The resilience and creativity of other researchers throughout Africa (Chiramba and Maringe, 2022; Makoe, 2022; Panebianco et al., 2023; Sayed et al., 2021; Shaik et al., 2022) inspired us. Thus, we embarked on the journey of compiling this edited book, which focuses on *STEM Education in the Post-pandemic Learning Space: Digitilization in Africa*. As discussed in the key points earlier in this chapter, lecturers throughout Africa developed innovative methods and techniques to support their digitalized pedagogy. From the infusion of technology and online devices to blended pedagogy, we have successfully navigated the digitalized STEM educational environment. In many African institutions, post-COVID-19, the curriculum needs reconfiguring to support transformative STEM pedagogy. Digital transformation and curriculum reconfiguration at many African education institutions required the development of digital and virtual spaces based on strengths, challenges, and opportunities that emerged (Mare et al., 2022; Mhlongo and Dlamini, 2022). These were founded on empirical evidence from the experiences of students and lecturers. The chapters in this book exhibit the resilience and adaptability of our students and lecturers. In-depth discussions, findings, and experiences are included in the chapters of this edited book.

Moreover, chapters in this edited book also focus on STEM teaching and learning in the new normal after COVID-19. Thus, based on the discussions in this book, we have learnt valuable lessons about adaptability, transformation, resilience and innovation during the COVID-19 pandemic. By implementing digital transformation thoughtfully and inclusively, we can successfully navigate our way to a sustainable future in STEM education across Africa (Ramnarain and Ndlovu, 2023; Shetye et al., 2022).

## Implications and Future Directions

The rapid digital transformation of STEM education, fast-tracked by the COVID-19 pandemic, presents essential implications for lecturers, educators, policy-makers, students, researchers, and the broader community (Bozkurt et al., 2020; Deroncele-Acosta et al., 2023). Understanding these implications is critical for using the possible benefits and opportunities and mitigating the challenges of these transformed educational environments.

Digitalization offers a unique opportunity to transform pedagogy, ensuring that education is more student-focused and interactive (Deroncele-Acosta et al., 2023; Islam et al., 2022; Megri et al., 2021). Lecturers and educators must welcome continuous professional development in digital literacy to successfully incorporate new technologies into their practice. Incorporating new technologies into the STEM curriculum can considerably enhance and support student engagement and learning outcomes. Thus, developing or revising the existing curricula that are adaptive and pertinent to the digital age is important for preparing students for the future. To ensure that the digital landscape is successfully navigated, support from institutions is vital (Barrot and Acomular, 2022; Chiramba and Maringe, 2022; Makoe, 2022). Accordingly, equal access to training programs, resources, and collaborative platforms can empower lecturers and educators to be innovative and successful using digital STEM pedagogy.

Consequently, policies need to be revised or developed to reconfigure the curriculum and have equal access to key infrastructure, training programs, devices, and resources. Thus, policy-makers play an important role in supporting and promoting digitalization in Africa (Vyas-Doorgapersad, 2022; Xulu, 2024). Policies need to focus on the digital divide to ensure equal access to quality education, particularly for marginalized students and students living in underserved areas. Moreover, encouraging partnerships with industry, government, and the private sector is important. These partnerships are important for ensuring access to internet connectivity and devices to ensure sustainable educational initiatives. Learning opportunities for all students need to be expanded so that all students are offered equal opportunities to succeed in a digital STEM educational environment (Siregar et al., 2023; Yang and Baldwin, 2020). Expanded learning opportunities will scaffold students as their critical thinking, creativity, digital literacy skills, and deeper understanding of STEM subjects are promoted. Furthermore, by encouraging a learning culture and providing support, the community plays an important role in connecting the gap between educational institutions and the resources they need.

Researchers also play an essential role in sustaining digital STEM in Africa (Chisom et al., 2023; Kanu and Anaekwe, 2022; Nwokocha and Legg-Jack, 2024). Further research is important to understand the long-term effects of digitalization on STEM education. Thus, ongoing research and innovation in pedagogy, educational models, and technologies are essential in identifying research gaps to guide future studies, inform policy decisions, and promote inclusive education systems.

## Conclusion

As is evident, the implications of digitalizing STEM education in Africa are extensive. Understanding and addressing these implications can create a more inclusive, innovative, and resilient educational landscape. Educators, policy-makers, students, researchers, industry, and communities must work together to harness the potential of digital learning and ensure that no student is left behind. As we look towards the future of STEM education in Africa post-pandemic, it is essential to build on the lessons learnt and take advantage of the opportunities presented by digitalization. Moreover, in Africa, the future of STEM education holds great potential, but it requires collaborative efforts from all stakeholders. We can create a more inclusive, resilient, and innovative educational environment by focusing on technological advancement, curriculum innovation, supportive policies, professional development, equity, research, student empowerment, and community engagement. Accordingly, educators, policy-makers, students, researchers, industry, communities, and other stakeholders must take practical steps to achieve this idea.

## Bibliography

Badat, S. (2020). Reproduction, transformation and public South African higher education during and beyond COVID-19. *Transformation: Critical Perspectives on Southern Africa 104*(1), 24–42. <https://muse.jhu.edu/article/777044>.

Barrot, J. S., & Acomular, D. R. (2022). How university teachers navigate social networking sites fully online: Provisional views from a developing nation. *International Journal of Educational Technology in Higher Education 19*(51), 1–19. <https://doi.org/10.1186/s41239-022-00357-3>.

Bozkurt, A., Jung, I., Xiao, J., Vladimirschi, V., Schuwer, R., Egorov, G., …, & Paskevicius, M. (2020). A global outlook to the interruption of education due to COVID-19 pandemic: Navigating in a time of uncertainty and crisis. *Asian Journal of Distance Education 15*(1), 1–126. <http://www.asianjde.com/ojs/index.php/AsianJDE/article/view/462>.

Caldararo, N. L. (2020). COVID-19, isolation, mammalian dispersal patterns, urban density, social distancing and mass psychogenic disease. *Isolation, Mammalian Dispersal Patterns, Urban Density, Social Distancing and Mass Psychogenic Disease* (May 1, 2020).

Chiramba, O., & Maringe, F. (2022). Organisational resilience as an urgent strategic goal in post-COVID-19 higher education in South Africa. In *Re-imagining Educational Futures in Developing Countries: Lessons from Global Health Crises* (pp. 39–63). Cham: Springer International Publishing. <https://doi.org/10.1007/978-3-030-88234-1_3>.

Chisom, O. N., Unachukwu, C. C., & Osawaru, B. (2023). STEM education advancements in Nigeria: a comprehensive review. *International Journal of Applied Research in Social Sciences 5*(10), 614–636. <https://www.fepbl.com/index.php/ijarss>.

Deroncele-Acosta, A., Palacios-Núñez, M. L., & Toribio-López, A. (2023). Digital transformation and technological innovation on higher education post-COVID-19. *Sustainability 15*(3), 1–24. <https://doi.org/10.3390/su15032466>.

Graham, C., Chun, Y., Hamilton, B., Roll, S., Ross, W., & Grinstein-Weiss, M. (2022). Coping with COVID-19: Differences in hope, resilience, and mental well-being across US racial groups. *PloS one 17*(5), 1–19. <https://doi.org/10.1371/journal.pone.0267583>.

Islam, M. K., Sarker, M. F. H., & Islam, M. S. (2022). Promoting student-centred blended learning in higher education: A model. *E-Learning and Digital Media 19*(1), 36–54. <https://doi.org/10.1177/20427530211027721>.

Kanu, A. C., & Anaekwe, M. C. (2022). Containing the impacts of the COVID-19 pandemic: A step towards sustaining basic scientific and technological skill acquisition in the society. *Stem Journal of Anambra Stan 4*(1), 30–41. <https://anambrastan.org/journals/index.php/stemjas/article/view/27>.

Knowles, J. R., Gray, N. S., O'Connor, C., Pink, J., Simkiss, N. J., & Snowden, R. J. (2022). The role of hope and resilience in protecting against suicidal thoughts and behaviors during the COVID-19 pandemic. *Archives of Suicide Research 26*(3), 1487–1504. <https://doi.org/10.1080/13811118.2021.1923599>.

Kolog, E. A., Egala, S. B., Amponsah, R., Devine, S. N. O., & Sutinen, E. (2022). COVID-19 pandemic: How can the lessons learnt contribute to the digital transformation of schools of tomorrow? *International Journal of Technology Enhanced Learning 14*(2), 142–162. <https://doi.org/10.1504/IJTEL.2022.121814>.

Maarman, G. J. (2023). Basic sciences in higher education, and teaching approaches in the context of 21st-century advances: time for a change? *South African Journal of Higher Education 37*(2), 132–150. <https://doi.org/10.20853/37-2-5016>.

Makoe, M. (2022). Resilient leadership in time of crisis in distance education institutions in Sub-Saharan Africa. In *Handbook of Open, Distance and Digital Education* (pp. 1–15). Singapore: Springer Nature Singapore. <https://doi.org/10.1007/978-981-19-0351-9_30-1>.

Mare, A., Woyo, E., & Amadhila, E. M. (2022). Harnessing the technological dividends in African higher education institutions during and post-COVID-19 pandemic. In *Teaching and Learning with Digital Technologies in African Higher Education Institutions* (pp. 1–24). Routledge. <https://doi.org/10.4324/9781003264026-1>.

Megri, A. C., Hamoush, S., Megri, I. Z., & Yu, Y. (2021). Advanced manufacturing online STEM education pipeline for early-college and high school students. *Journal of Online Engineering Education 12*(2), 1–6. <https://www.onlineengineeringeducation.com/index.php/joee/article/view/47>.

Mhlanga, D., Denhere, V., & Moloi, T. (2022). COVID-19 and the key digital transformation lessons for higher education institutions in South Africa. *Education Sciences 12*(7), 1–17. <https://doi.org/10.3390/educsci12070464>.

Mhlongo, S., & Dlamini, R. (2022). Digital inequities and societal context: Digital transformation as a conduit to achieve social and epistemic justice: Digital transformation as a conduit to achieve social and epistemic justice. In *Innovation Practices for Digital Transformation in the Global South: IFIP WG 13.8, 9.4, Invited Selection* (pp. 1–15). Cham: Springer International Publishing. <https://doi.org/10.1007/978-3-031-12825-7_1>.

Naidoo, J. (2020). Postgraduate mathematics education students' experiences of using digital platforms for learning within the COVID-19 pandemic era. *Pythagoras 41*(1), 1–11. <https://doi.org/10.4102/pythagoras.v41i1.568>.

Nwokocha, G. C., & Legg-Jack, D. (2024). Re-imagining STEM education in South Africa: leveraging indigenous knowledge systems through the m-know model for curriculum enhancement. *International Journal of Social Science Research and Review 7*(2), 173–189. <http://dx.doi.org/10.47814/ijssrr.v7i2.1951>.

Panebianco, C., Staden, W. V., Lotter, C., & Plessis, R. D. (2023). Resilience, art activities, and income of artists in a low-to-middle-income country during the COVID-19 pandemic. *Global Health Econ Sustain 1*(1), 1–12. <https://doi.org/10.36922/ghes.0911>.

Ramnarain, U., & Ndlovu, M. (2023). Reflections and future directions for ICT integration in STEM education for Africa. In *Information and*

*Communications Technology in STEM Education* (pp. 215–221). Routledge. <https://doi.org/10.4324/9781003279310>.

Sayed, Y., Singh, M., Bulgrin, E., Henry, M., Williams, D., Metcalfe, M., …, & Mindano, G. (2021). Teacher support, preparedness and resilience during times of crises and uncertainty: COVID-19 and education in the Global South. *Journal of Education* 84(1), 125–154. <http://dx.doi.org/10.17159/2520-9868/i84a07>.

Sedaghatjou, M., Hughes, J., Liu, M., Ferrara, F., Howard, J., & Mammana, M. F. (2023). Teaching STEM online at the tertiary level during the COVID-19 pandemic. *International Journal of Mathematical Education in Science and Technology* 54(3), 365–381. <https://doi.org/10.1080/0020739X.2021.1954251>.

Shaik, N., Dippenaar, A., Kwenda, C., Petersen, K., Esau, D., & Oliver, H. S. (2022). Sink or swim: Exploring resilience of academics at an education faculty during Covid-19. *Journal of Education* 89(1), 169–185. <http://dx.doi.org/10.17159/2520-9868/i89a09>.

Shetye, N., Lotz-Sisitka, H., Albrecht, E., Durr, S., Marx, D., Chirambo, D., …, & van Zyl-Bulitta, V. (2022). Digitilization and transformative learning for sustainable futures in rural Africa: Leaving no one behind. In *Africa–Europe Cooperation and Digital Transformation* (pp. 199–214). Routledge. <https://doi.org/10.4324/9781003274322>.

Siregar, N. C., Gumilar, A., Warsito, W., Amarullah, A., & Rosli, R. (2023). Enhancing STEM learning for all: A paper concept of accessible resources. In *International Journal of Applied Sciences and Sustainability* 1(1), 58–68. <https://ejournal.uika-bogor.ac.id/index.php/IIJASS/article/view/14309>.

Vyas-Doorgapersad, S. (2022). The use of digitilization (ICTs) in achieving sustainable development goals. *Global Journal of Emerging Market Economies* 14(2), 265–278. <https://doi.org/10.1177/09749101211067295>.

Xulu, Z. C. (2024). Integrating technology in teaching African languages in South African universities: A call for digitilization. *Journal of the Digital Humanities Association of Southern Africa* 5(1), 1–14. <https://doi.org/10.55492/dhasa.v5i1.5030>.

Yang, D., & Baldwin, S. J. (2020). Using technology to support student learning in an integrated STEM learning environment. *International Journal of Technology in Education and Science* 4(1), 1–11. <https://scholarworks.boisestate.edu/edtech_facpubs/239/>.

# About the Editors

**Brantina Chirinda** is a senior research associate at the University of Johannesburg and a postdoctoral research fellow at the University of California, Berkeley. She holds a Ph.D. in Mathematics Education from the University of Witwatersrand. Her primary research interests are in the teaching and learning of mathematics in contexts of disadvantage, specifically focusing on mathematical problem-solving and equitable access to content in the mathematics classroom. For over 20 years, she has taught Mathematics and Mathematics Education courses at various Southern African institutions and the United States of America. She has published several books, conference proceedings, book chapters, and articles in accredited local and international journals. She has presented her work at several national, regional, and international conferences. She leads a community project, Southern African Mathematics Empowerment Network (SAMEN), which is envisaged to empower female mathematics teachers and learners in contexts of disadvantage.

**Jayaluxmi Naidoo** is an Associate Professor of Mathematics Education at the University of KwaZulu-Natal (UKZN) Edgewood Campus. She received the UKZN Emerging Researcher Award (2012) and was successful in obtaining both the SANPAD (2009) and Canon Collins scholarship (2005). In 2022, she received the Top 3 Most Published Researcher award from the School of Education. In 2018, she received the Top 30 Most Published Researcher and Top 10 Most Published Women Researcher awards from the College of Humanities, UKZN. She is an NRF-rated researcher, and in 2021, she published an edited book on teaching and learning in the twenty-first century, embracing the Fourth Industrial Revolution (Brill Publishers). Her research interests include the use of technology in STEM education, visualization in mathematics, issues of social justice, race, language, and STEM education, pedagogic strategies to improve the teaching and learning of STEM subjects, indigenous knowledge systems and STEM education, and mathematics teacher training and development.

# Notes on Contributors

**Crispen Bhukuvhani** holds a Doctor of Philosophy (D.Phil.) in Chemistry Education. He is the Executive Dean of Research Innovation, and Postgraduate Studies at Manicaland State University of Applied Sciences, Zimbabwe. He has extensive experience in science and technology education and research interests in Science, Technology, and Innovation education ecosystems.

**Locadia Bhukuvhani** holds a Master of Education in Educational Psychology and has experience as an educator in the Zimbabwe and Rwanda education systems. She is a Lecturer in Educational Foundations—Psychology of Education at the Rubengera Teacher Training College in Karongi, Rwanda.

**Charles Chikunda** (Ph.D.) coordinates the UNESCO Education for Sustainable Development (ESD) and Global Citizenship Education (GCED) programs. He has extensive experience in teacher education, going beyond 20 years in the SADC region. He has also worked in development work, supporting capacity development in communities, civil society organizations, and governance structures in natural resources management.

**Alois Solomon Chiromo** is a professor and the Executive Director of Quality Assurance and Professional Development at Midlands State University, Zimbabwe. He also served as the Acting Pro-Vice-Chancellor for Research and Academic Affairs and Executive Dean of the Faculty of Education at the same university.

**Nosisi N. Feza** is the Deputy Vice Chancellor of Research and Postgraduate Studies at the University of Venda (Univen) in Thohoyandou. Before she joined UNIVEN, she was the Rector of the Buffalo City Campus at Walter Sisulu

University. She has also served as a Dean at the Central University of Technology, coming from the University of South Africa as the Head of an Institute for Science and Technology Education. She has also worked as a Senior Research Specialist at the Human Sciences Research Council. She has been actively involved in teaching, learning, and research in her previous institutions of employment: the University of Northern Iowa in the US as a Research Fellow, the State University of New York as a research assistant, and Nelson Mandela University, focusing on mathematics education. She is a C-rated researcher by the National Research Foundation. She is an Awardee of a lifetime achiever from Nelson Mandela University and a STEM leader awardee of HERS Academy.

**Rajendran Govender** is currently the Dean of the Faculty of Education at the University of the Western Cape and also served as Deputy Dean of Teaching and Learning in the Faculty of Education from 2017 to 2020. His area of mathematics education research focuses on Geometry, problem-solving, modeling, reasoning and proof, and ICTs to facilitate meaningful learning. He is currently the Editor-in-Chief of the Pythagoras journal. He has recently been appointed as a participating CHE Initial Teacher Education Reference Group member and the National Institute for Curriculum and Professional Development (NICPD) Online Teacher Development Platform Reference Group member. At the University of Western Cape, he recently served as chairperson of the IOP (2021–2025) task team 3 for Goal Area 2—Learning and Teaching. He has coordinated the development of the new Assessment Policy for UWC in 2020–2021.

**Neliswa Gqoli** is a Senior Lecturer at Walter Sisulu University (WSU) in the Department of Adult and Educational Foundations, where she teaches Psychology of Learning and Mathematics Teaching. She obtained her master's degree at Walter Sisulu University and Ph.D. (Early Childhood Development) at the University of Free State. Her research focus is on Early Childhood Development and Mathematics Education. She is a reviewer for the Interdisciplinary Journal of Sociality Studies and the Cogent Education Journal. As the Senior Researcher in the Faculty, she has published articles and book chapters nationally and internationally. She taught mathematics for 25 years in Eastern Cape Primary Schools. She is a Faculty Board Member and a coordinator for the Postgraduate Certificate in Education at WSU. She is also the chairperson of the Faculty Social Committee. She is a South African Research Association for Early Childhood Education (SARAECE) member.

**Zingiswa Jojo** is a full professor of mathematics education in the Department of Secondary and Post-School Education at Rhodes University. She serves on the

Commission for African Women in Mathematics (CAWM) committee (South African Chapter). Her primary research interests include the teaching and learning of geometric concepts, Conceptual understanding of Calculus Concepts, Instructional design in mathematics teaching, Ethno mathematics and Indigenous knowledge, Mathematics in-service teacher and pre-service professional development and Material Development for Mathematics Education, including the use of Open Education Resources that are relevant to the teaching of mathematics at school and tertiary levels.

**Esther Samwel Kibga** holds a Ph.D. in Chemistry Education from ACEITLMS of the University of Rwanda, College of Education-Rukara campus. She is a lecturer in Science Education at the Aga Khan University AKU-IED P.O. Box 125, Dar es Salaam. Her research areas of interest are STEM education, climate and environmental sustainability, teacher education, gender in science education (STEM), assessment and evaluation, early childhood education, primary education, e-learning, and creating a transformative future generation that reflects the technologically changing world. Her career aspiration is to become a science education and research specialist.

**Mariam Makramalla** is an assistant professor at the University of NewGiza, Egypt. She holds a Ph.D. in Mathematics Education from Cambridge University. She is a critical scholar whose thinking is grounded in political economy and decomposing curriculum reform to ask whether reforms serve to reproduce colonialist thought. In this vein, she investigates sociocultural power dynamics and politics, anti-colonialist thought, and (perceived lack of cultural/communal) assets in education. One of her particular areas of interest relates to the material differences in education in Africa versus the West, how historical political arrangements create situations of inequity, and the implications for this in the cross-cultural exchange of ideas on curriculum reform.

**Puleng Motseki** holds a Doctoral degree in Mathematics Education from the University of Johannesburg (2021) and a master's in Education from the University of Johannesburg (2019). Her current research interest is how Technical Vocational Education and Training students understand differential calculus concepts.

**Fredrick Mtenzi** is a faculty member and an associate professor at the Aga Khan University, Institute for Educational Development—East Africa. His research interests range from designing secure mobile financial transactions in low-end mobile devices to security auditing and developing de-identification strategies, which increase the utility and privacy of sensitive information in the era of big data. Recently, he has been involved in researching ways of ensuring software development is cognizant of different learning styles and effective ways of scaling up the utilization of ICT with minimal instructor intervention. He is also involved in two national projects to improve the quality of teachers in Tanzania by providing continuous professional development for teachers using affordable technology.

**Eddie M. Mulenga** holds an earned PhD in Mathematics Education from the University of Valladolid, Spain. He is currently a postdoctoral research fellow at the University of Johannesburg. He is a published author of various peer-reviewed papers in top accredited international journals. He has attended and presented papers at various local and international conferences in Mathematics Education. With his emphasis on the usage of digital tools, he has taught Mathematics and Mathematics Education courses at various Zambian institutions for over seven years.

**Shonisani Agnes Mulovhedz**i holds a Ph.D. in Early Childhood Education from the University of Pretoria. She is a committee member of South African Research in Early Childhood Education (SARAECE) and a provincial early childhood development inter-sectoral member. She is a Univen Senate member. She is also a coordinator of the Univen-Model Preschool. Her research and teaching relating to Early Childhood Education emphasize leadership, Life Skills, Mathematics education, and inclusive education in the Foundation Phase. She won Vice-Chancellors Excellence awards in Research, teaching, and Learning. She has written and published book chapters and articles in accredited journals. She has authored and co-authored numerous articles and book chapters. She moderated dissertations and theses from various universities and reviewed articles, books, and book chapters. She is HOD in the Department of Early Childhood Education at the University of Venda, and she is a C3-rated researcher.

**Agnes Pakombwele** is a lecturer in the Department of Basic Education and Childhood at Baisago University, Gaborone Campus, Botswana. She holds a Ph.D. in Early Childhood Development and Learning from the University of the Free State. She also holds a Master of Education Degree in Early Childhood Development and a Master of Social Sciences in Child and Family Studies. Her research interests include STEAM teaching and learning in early years, early childhood teaching pedagogies in the twenty-first century, early childhood care and education, and early childhood policy and practices.

**Asheena Singh-Pillay** is an Associate Professor of Technology Education at the School of Education, College of Humanities, University of KwaZulu-Natal. She is also the Academic Leader for the Bachelor of Education (B.Ed.) program at the School of Education, University of KwaZulu-Natal (UKZN). She has presented her research at conferences nationally and internationally. In 2018, she received an award for the Top 30 Most Published Researcher and a Top 10 Most Published Women Researcher award from the College of Humanities, UKZN. Her research interests include design process, spatial visualization thinking and skills, engineering graphics and design, socially responsive STEM education, gender and education, and sustainable development.

**Tawanda Runhare** is a Professor and Director of the School of Education at the University of Venda, South Africa. He holds a Ph.D. in education from the University of Pretoria, South Africa. He has published widely and presented conference papers on issues that influence educational access, participation, and outcomes. He is the lead editor of the book Sociological Foundations of Education in Africa, published in 2021 by Oxford University Press Southern Africa. His research is founded on vast experience in education as a teacher and school principal, regional and national chief examiner, teachers' college senior lecturer, university lecturer, senior lecturer, associate professor, professor, head of department, school of education dean, and director.

**Gladys Sunzuma** is a senior lecturer in the Science and Mathematics Education Department at the Bindura University of Science Education. She holds a Ph.D. in mathematics education from the University of Kwa-Zulu Natal. She has served for 7 years as a high school mathematics, physical science, integrated science, and chemistry teacher in Zimbabwe, 3 years as a teaching assistant in the Education Department at Bindura University of Science Education, and over 12 years in her current position: senior lecturer at the Bindura University of Science Education. She has published over 35 articles in various local and international journals. She had reviewed several articles from various journals. Her research interests include finding ways to improve the teaching and learning of mathematics both at the school and university levels and improving teacher training programs to include better pedagogical content knowledge for teachers.

# Africa in the Global Space

## Edited by Edward Shizha

Africa in the Global Space is an innovative and scholarly space providing analysis and interrogation of diverse perspectives on Africa's role, and contributions to global socio-cultural, political, educational and developmental debates. The series provides an-up-to-date scholarly appraisal of research on various subjects, including issues of paramount importance to globalization and development in Africa (politics, democracy, education, economics, philosophy, religion, gender, technology, global relationships and the role of government and non-governmental organizations). The series is dedicated to increasing understanding of Africa's internal and international relations, and developmental trends and policies, through comparative, cross-cultural and international perspectives. Developed by an international editorial board of emerging and established scholars, this series is a visionary and interdisciplinary space to engage and inform debate on Africa's participation in the global nexus.

| | |
|---|---|
| Volume 1 | Kenneth Mahuni, Josiah Taru and Wellington G. Bonga. Africa's Incomplete Cycles of Development. 2020. |
| Volume 2 | Mishack T. Gumbo. Decolonization of Technology Education: African Indigenous Perspectives. 2020. |
| Volume 3 | Christopher Ndlovu and Edward Shizha. The Dynamics of African Indigenous Knowledge Systems: A Sustainable Alternative for Livelihoods in Southern Africa. 2022. |
| Volume 4 | Remi Prospero Fonka. The Nso' Concept of Time: An African Cosmological Perspective. 2023. |
| Volume 5 | Yemi Ogunyemi. Yoruba Idealism. 2022. |
| Volume 6 | Brantina Chirinda, Lwazi Sibanda et al. Science, Mathematics, and Technology Education in Zimbabwe: Research, Policy and Practice. 2023. |
| Volume 7 | Cyriaque Sobtafo. The Effectiveness of Official Development Assistance in the Health Sector in Africa: A Case Study of Uganda. 2023. |
| Volume 8 | Moha Ennaji. Les nouveaux défis de la migration Maghreb-Europe. 2024. |
| Volume 9 | Wilson Dabuo. A History of the People of Nandom, 1660–1955. 2025. |
| Volume 10 | Megan Goins. A Hidden Genocide: The Profitable War in the Democratic Republic of the Congo. 2025 |
| Volume 11 | Brantina Chirinda and Jayaluxmi Naidoo (eds.). STEM Education in the Post-Pandemic Learning Space: Digitilization in Africa. 2026. |